HARVEST

Also by Richard Horan

FICTION

Goose Music
Life in the Rainbow

NONFICTION

Seeds: One Man's Serendipitous Journey to Find the Trees That Inspired Famous American Writers

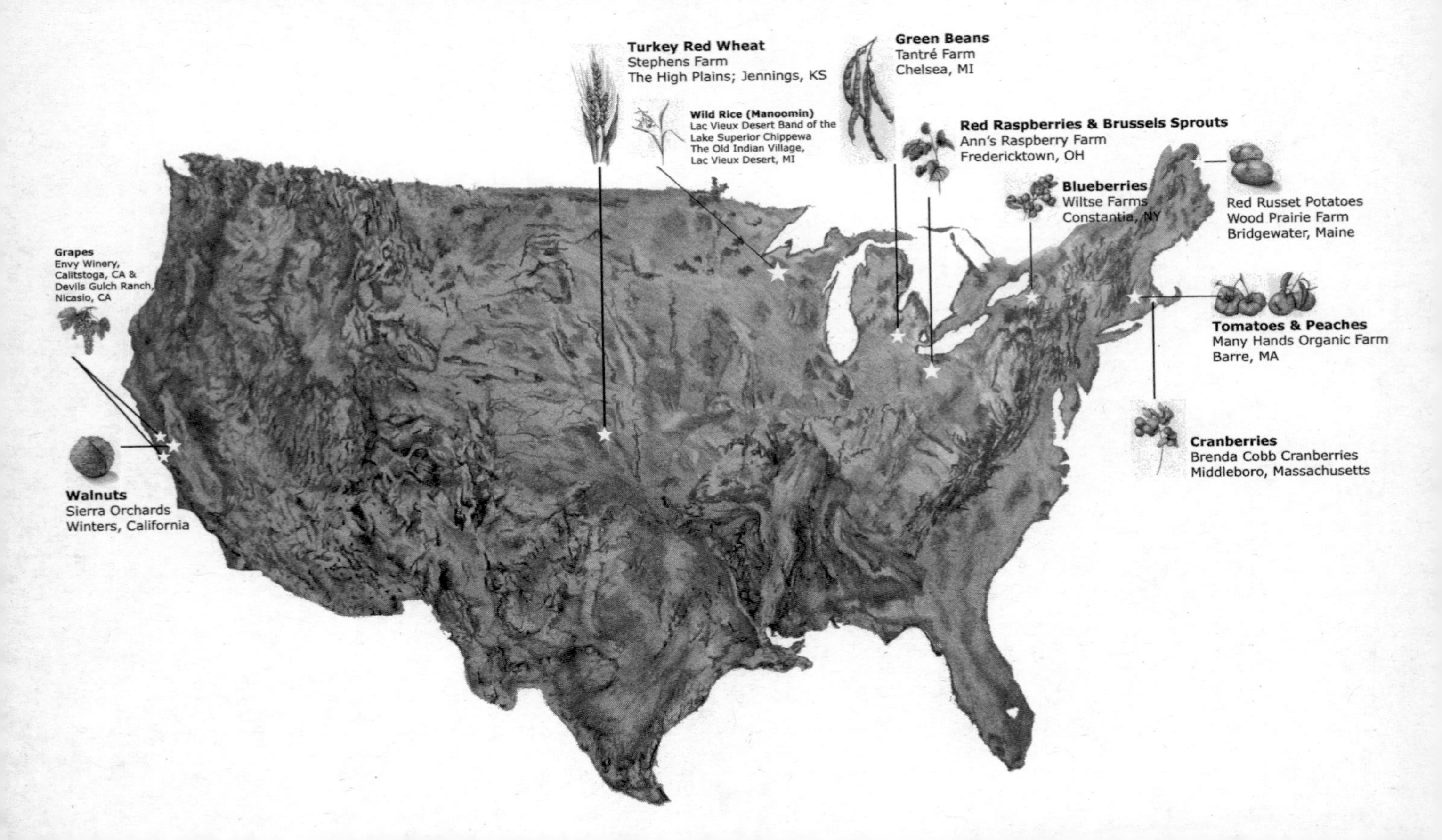

Grapes
Envy Winery, Calitstoga, CA & Devils Gulch Ranch, Nicasio, CA
Walnuts
Sierra Orchards
Winters, California
Turkey Red Wheat
Stephens Farm
The High Plains; Jennings, KS
Wild Rice (Manoomin)
Lac Vieux Desert Band of the Lake Superior Chippewa
The Old Indian Village, Lac Vieux Desert, MI
Green Beans
Tantré Farm
Chelsea, MI
Red Raspberries & Brussels Sprouts
Ann's Raspberry Farm
Fredericktown, OH
Blueberries
Wiltse Farms
Constantia, NY
Red Russet Potatoes
Wood Prairie Farm
Bridgewater, Maine
Tomatoes & Peaches
Many Hands Organic Farm
Barre, MA
Cranberries
Brenda Cobb Cranberries
Middleboro, Massachusetts

HARVEST

An Adventure into the Heart of America's Family Farms

RICHARD HORAN

HARPER PERENNIAL

NEW YORK • LONDON • TORONTO • SYDNEY • NEW DELHI • AUCKLAND

Illustrations by Debbie Ferrer.

Map on p. iv by Matthew Ferrer.

FIRST EDITION

Designed by Janet M. Evans

Library of Congress Cataloging-in-Publication Data is available upon request.

ISBN 978-0-06-209031-7

12 13 14 15 16 OV/RRD 10 9 8 7 6 5 4 3 2 1

For my joyous and musical wife, Mary, who, after harvesting cranberries all day long and while trying to enjoy a plate of fried clams at a seaside restaurant in Plymouth, was instead subjected to listening to "Uptown Girl" by Billy Joel, "Wildfire" by Michael Martin Murphey, Kenny Loggins's "Danger Zone," Pat Benatar's "Hit Me with Your Best Shot," Jackson Browne's "Doctor My Eyes," and, last but not least, David Soul caterwauling "Don't Give Up on Us." She turned a sour note into sweet harmony by confiding: "You know, I am so fortunate to have music in my life—to be able to play music and perform it and yet never have to take it as seriously as Jackson Browne."

And for all of the family farmers and farmhands who are America's national treasures.

"Is it too outlandish to wonder if, as a new agricultural economy becomes more like nature's economy, our minds will expand?"

—WES JACKSON,
CONSULTING THE GENIUS OF THE PLACE

"The most insistent and formidable concern of agriculture, wherever it is taken seriously, is the distinct individuality of every farm, every field on every farm, every farm family, and every creature on every farm."

—WENDELL BERRY,
THE WAY OF IGNORANCE

CONTENTS

Introduction

POTTERSVILLE

Midsummer 2010. Oswego, New York. I am sitting in my rusted-out Dodge Grand Caravan, in the soporific parking lot of the local Wal-Mart, waiting for my teenage daughter, Evelyn, to purchase a new cell phone. It is a belated birthday gift, but with no steady income, it is not one I can easily afford. I am listening to the radio. The president of the United Farm Workers, Arturo Rodriguez, is being interviewed on NPR about his "Take Our

Jobs" campaign.* He is talking about how the United Farm Workers has issued a challenge to all Americans, especially those who complain that undocumented workers are crossing the border and stealing *their* jobs: "Apply to become a farm worker!"† He has just revealed that up until now, only about four thousand people had filled out applications to become farmworkers, and from those only a few dozen have followed through to the end. I am sweating and feeling a little sick to my stomach because below my feet the newly tarred surface of the parking lot is exuding a toxic vapor that has got to be mutating my chromosomes, killing my brain cells, and doing who knows what to my sperm count. The interview comes to an end and Rodriguez sums it all up with characteristic verve: "So it's a grueling effort, a grueling job that takes place and they get very little recognition for what they do. But the reality is that if it wasn't for them, we would not have food on our tables every single day."

I turn off the radio and sit there considering this truth, while at the same time watching the parade of the morbidly obese shuffling into the cardboard-cutout box store, not a tree or a blade of grass or a piece of fruit within a thousand feet. Suddenly

---

* "'Take My Job!' Campaign Markets Agricultural Labor," July 7, 2010, NPR.

† On its website TakeOurJobs.org, the UFW generously offered to "train citizens and legal residents who wish to replace immigrants in the fields," along with the asterisked, tongue-in-cheek caveat that the "[j]ob may include using hand tools such as knives, hoes, shovels, etc. Duties may include tilling the soil, transplanting, weeding, thinning, picking, cutting, sorting & packing of harvested produce. May set up & operate irrigation equip. Work is performed outside in all weather conditions (Summertime 90+ degree weather) & is physically demanding requiring workers to bend, stoop, lift & carry up to 50 lbs on a regular basis."

I feel disoriented, lost, discombobulated . . . *Where am I? Who am I? What's happening here?*

My stomach lurches. I close my eyes, press the balls of my feet against the floorboard, and drift back . . . back . . . back into my childhood . . .

*Midsummer 1969. I'm in my grandfather's refrigerated warehouse, an old brick building on Spruce Street in the Italian section of Stamford, Connecticut. He is the owner of a produce wholesale business, John Vitti, Purveyor of Fine Fruits and Vegetables. I'm sitting on a wooden packing crate next to a large oak desk, and swinging my legs. To my right is a yawning, garage-style door through which people of an odd assortment of ethnicities continually enter and exit. I watch an ancient Italian woman shuffle in wearing a black-widow dress and a black shawl over her steel-gray head of hair. I watch her amble, birdlike, to the back of the store and rummage through the dozens of wax boxes heaped full of fruits and vegetables. There are other women rummaging through the boxes back there, too: mostly Italians, but some Puerto Ricans and Poles. Nearby, my grandfather, a bald, bull-necked man, stands patron-like among a group of short men, all in neatly ironed short-sleeved shirts, nodding and gesticulating, and speaking in Italian. They are extolling the virtues of the produce—*pomodori, prugne, cocomeri, meloni tutti quanti sono buoni*—one of the best harvest seasons ever. The sun is dazzling bright outside on the street, casting short, squat shadows. Inside, the cool moist air is permeated with the piquant odor of ripe fruit and fresh vegetables with just a hint of putrefaction. Everything is alive and lush. The black widow comes up to the* *desk, her canvas bag full of peaches. I stop swinging my legs and look up into her eyes. They are deeply set and the color of burnt almonds. She waits for my grandfather to come over and count her out. She looks at me and begins to rub my face with her gnarled hand. Her skin is dry as paper, but her toothless smile is an explosion of joy and sun-dried peace. There is a brown mole on her chin the size and shape of a chocolate raisin. She hands me a peach. I can tell she wants to vicariously eat it by watching me do so with my nearly complete set of teeth. I bite into it. The sticky juice tickles my cheeks as it runs down my face. She stands back shaking her head with great satisfaction, then motions with both hands, imploring my grandfather to look:*

"Guarda! Guarda!" she beams.

He glances down at me, smiles, hands her the change, and then lovingly rubs my head: "Bel ragazzino, eh?"

"Non ragaz, la pesca! Che succos'!" (Not the kid; the peach! How juicy!)

I am suddenly knocked headlong out of my reverie, my stomach dropping down into the bottom of my belly. My daughter plops excitedly down in the passenger seat beside me. She holds the new device in front of my face so I can admire it: "Look! Look!" It is square, black, odorless, and dead. I look up at her face; her smile is not as ripe, or bright, or nearly as fully satisfied as the black widow's. "Dad, what's wrong?" she asks as I continue to stare at her, my eyes tearing up, my facial expression dripping with despair.

"Nothing. Nothing," I lie.

But in fact, dear reader, everything at that moment was wrong. *Everything!* To start with, I was unemployed. Actually, I just lied again: it was worse than that—I was working as a substitute teacher at the local high school during the school year. Plus, to make ends meet, my wife and I were renting out rooms in our home and sharing our kitchen with four Chinese students. Have you ever tried to share a kitchen with four Chinese women? It's like trying to toast four slices of bread in a two-slot toaster. Make that six slices of toast. Added to that, my oldest daughter had quit college after two years because she felt like it was a waste of time and money, *and she was right*. Meanwhile, all across the nation, the ash trees were dying from the emerald ash borer. Biodiversity was declining at an unprecedented rate. Global warming was melting the polar ice caps. Entire island nations were sinking into the sea. The BP oil spill kept growing. Ocean gyres were filling with plastic. The population was increasing. The gap between the rich and the poor was growing wider. There was nuclear contamination. Haiti. Tibet. Afghanistan. Fucking Arizona! Health-care costs. Tuition. Peak oil. Fracking. Lobbyists. Corruption. Derivatives. Credit default swaps. Goldman Sachs. Glass-Steagall? Citizens United! Corporate tax loopholes. Corporate personhood laws. Corporate monopolies. Hell, we were all living in Pottersville!

Over the course of the next few weeks, the images of the Italian black widow, her smile, the little Raisinet on her chin, the juicy peach, the taste of the peach, the smell of the fruits and vegetables, my grandfather's freshly washed and pressed shirt, all of them, kept popping into my mind; plus, the sound of Arturo

Rodriguez's voice, his singing optimism about the farmworkers, the grueling work, the purity of the work, the beloved workers, the sweet harvest . . . the harvest . . . kept ringing in my ears.

Then one morning I awoke with a singular notion. I wrote it down on a piece of paper and shared it with my wife. She liked it. I called my literary agent. She liked it, too. So, over the course of the next several weeks, we put a proposal together, and before you know it I had a book contract in hand. It was a very simple concept. It went like this: I would travel around the country—East, West, North, South—and participate in the harvests of a dozen crops. No fish. No meat. No dairy. Just fruits and vegetables, nuts and grains. Because I wanted it to be thoroughly and completely upbeat, we decided that I should not apply to become a United Farm Worker, where I would likely encounter exploitative and inhumane conditions that would force me to write yet another depressing muckraker like Upton Sinclair's *The Jungle* or Eric Schlosser's *Fast Food Nation* (or, a year later, Chris Hedges's "Tomatoes of Wrath.").* No, I would avoid that sort of thing altogether. Instead, I would stick to the small farms, the organic farms, the family-run farms, so that I could be sure to enjoy the fruits and vegetables of my labor.

It didn't take me long to put an itinerary together. I started by sending out solicitation letters† to farms where I could simply and

---

* Chris Hedges, "Tomatoes of Wrath in Immokalee, Florida," http://dandelionsalad.wordpress.com/2011/09/26/tomatoes-of-wrath-by-chris-hedges/.

† Dear Farmer:

I am planning a great American adventure. Throughout the coming year, I hope to travel across America from Maine to California, Florida to Washington, Wisconsin to Arizona in order to participate in the harvest happily participate in the harvest, nothing fancy or overly demanding. One by one, in their own special way, a dozen farms around the country invited me to participate—potatoes in Maine, wheat in Kansas, cranberries in Cape Cod, wild rice in the Upper Peninsula of Michigan, blueberries close to home, etc. There were some crops that I really wanted to pick, but they never materialized, no matter how many letters I wrote or phone calls I made or personal connections I pursued. Crops like peanuts, sweet potatoes, and oranges. Nevertheless, by the time the next harvest season rolled around, I had the imprimatur of a dozen farms.

The following pages describe my wonderful adventures and misadventures pitching in and helping out with the harvest. Come along with me, please do . . .

---

of some of America's most unique and beloved crops. I would very much like to participate in your annual harvest. I will not require anything other than the permission to chip in and help out as well as the okay to write about my experiences in a book and perhaps take a few photographs. The book will be called *Harvest*. This will be an upbeat, engaging, and hopefully inspiring narrative about the amazing food still available to us all notwithstanding the monopolistic pressure from the Monsantos, DelMontes, ADMs, etc. If this is something that you would be interested in helping me put together, please contact me at your earliest convenience. Thank you for your time and interest.

Sincerely,
Richard Horan, author

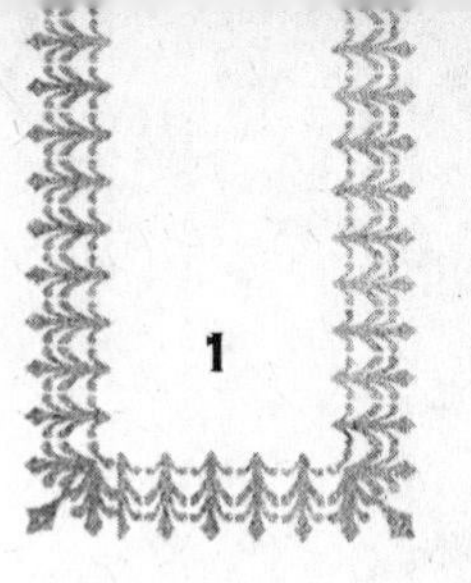

Turkey Red
Wheat

STEPHENS FARM

THE HIGH PLAINS; JENNINGS, KANSAS

JULY 2011

Due east out of Denver on Interstate 70. Driving toward the High Plains of Kansas, I felt like God was gently tipping me out of the palm of His hand down into the lap of Eden. On all four sides of me was endless green grass throbbing and pulsating with the photosynthetic power of the sun, all of which was swaddled in a vault of pure, pasteurized blue. I'd never before experienced an infinite horizon line like that, not from the mountaintops of Maine or from the shores of the Yucatán Peninsula. The High Plains of Kansas define freedom for the landlubber. The Cheyenne, the original inhabitants of the area, have a word for it: *toxto*, "open country/place of freedom."*

At Colby, I turned off I-70 and headed northwest toward Jennings. In no time I was motoring along a two-lane highway amid cornfields and cattle ranches. I was going about seventy-five miles per hour, yet even at that speed I felt like I was standing still. Occasionally a huge semi loaded with grain or cattle would appear in the far distance, coming from the opposite direction, and within moments would explode by my vehicle. And I mean explode. As the great monster trucks drew near, I felt the car hesitate and then sag, and an instant later—*WHAM*—the

---

* Linda Davis-Stephens (*Na ho'emaneveho'a*, Cheyenne given name), "Cheyenne Country Almanac of an Organic Farm Buffalo Ranch," paper presented July 28, 2009, for the International Union of Anthropological and Ethnological Sciences, in Kunming, China.

force of the semi's wake blasted against my vehicle, shimmying it sideways a foot or more. Then I noticed something odd—the rumble strips were not on the shoulders of the road, but in the middle of it! *I was in Kansas now.*

At the town of Jennings, a forlorn little cattle stop whose rusted, black-and-white street signs from the middle of the last century perfectly described and defined it, I called my host farmer, Bryce Stephens. He answered on the second ring.

"... take a right at the big silos—two miles east, three miles south, two and a half miles east again. We're in an old, two-story farmhouse. Any questions?"

"No road signs?"

"Nope."

As I drove slowly along the well-crowned dirt byways (there was no pavement, either), I kept one eye on the odometer and one eye on the amazing landscape. The place left an odd impression on me. On a stage of such sprawling magnitude, the distant trees and houses and animals all appeared disproportionately large and unambiguously anthropomorphic. That is to say, there was a sort of optical/psychological illusion happening that caused the objects in the distance to appear not only closer but also more psychically significant. It was as if the emotional center of my brain was filling in the disconcerting expanse with extra warm and fuzzy thoughts and items, compensating for the lack of close quarters that Mother Earth normally provides.

And then I was there. I had followed the coordinates as best I could without the aid of GPS, but I still wasn't a hundred percent sure that the driveway of the farmhouse I had pulled into was the one I wanted. But there was no turning back now, because within seconds a large pack of dogs had surrounded my car. I counted eight beasts of various sizes and breeds, two of which were eye level to the car window. They crowded around my side of the car, barking at my glass-enclosed image, pretending to menace with tails playfully wagging behind them. Being a close personal friend of *Canis lupus familiaris*, I was not hesitant to interact with them. But what really drew me out of the vehicle with alacrity was the sudden appearance of a large male human, who stood, among a fleet of out-of-date motor vehicles, hands on hips, feet shoulder width apart, baseball hat on head with steel-gray hair dripping out from under it. He had a great big chest, like he was wearing a chest protector. In fact, he looked just like an umpire who has stepped out from behind the plate to stare down a carping pitcher. From fifty feet I could make out a broad, carnivorous lower mandible. He had leg-of-mutton hands. His skin was dark as toasted wheat. And as I approached amid the amicable crowd of canines, all of whom were sniffing at my lower extremities and places in between, I could see two burning black coals under the shadow of his hat's brim. He looked upon closer inspection like a marine general or an Indian chief. Or both.

We shook hands. Yikes, what a grip!

He was a big man—six foot tall, two hundred and fifty pounds—and about sixty years old.

"You found the place okay?" He smiled.

"I just followed your directions."

Then, suddenly, from out behind one of the countless cars and trucks parked here and there in various conditions of functionality, a young woman appeared, in her late teens or early twenties. She wore a faded green safari hat, with tufts of brunette hair flowing out beneath, a reddish khaki shirt with the sleeves rolled up, a pair of faded jeans, and heavy work boots.

Her swanlike eyes were deep brown, and her skin was fair and taut; her lips and cheekbones were broad and high. She was well formed, muscular even, average height, painfully shy, and pretty as could be. But the most prepossessing quality about her was the dirt she wore—her clothes and her hands, even her face, were as grease-stained and dirty as any clothes and hands and face I had ever before laid eyes on. She was a real-life Dorothy Gale but wearing Huck Finn's clothes.

Her eyes looked straight up into mine as she introduced herself: "Demetria."

We shook.

"Demetria? Like the Greek goddess of the harvest?" I asked incredulous.

She smiled demurely, and then went about her business.

At that point Bryce clapped me on the back—ouch!—and led me inside the old homestead.

The house was small and dark and woefully cluttered compared with my obsessive-compulsive East Coast standards. Through a long, windowed passageway, full of boots and shoes and coats, we entered the living room. There was a kitchen off to the left and a bedroom and dining room and small bathroom straight ahead toward the back part of the first floor. A closed door to the right led upstairs to the second floor. Bryce's L-shaped work desk took up half the space in the living room. So in essence, the living room was also the office. There were books and papers scattered every which way on top of the desk. There was a big-screen television that separated the living room/office from the dining room, and innumerable Indian artifacts displayed on bookshelves and hanging from the walls. It suddenly hit me that Bryce himself was Native American. Not only did his gray hair and color suggest it, but also his mannerisms and his odd inflections—he talked like an Indian. I would soon learn that I was partly correct.

I took a seat in the large reclining chair to the left of his desk, or to the west of him, pausing for a moment to consider the significance of this western orientation—toward the setting sun and Rocky Mountains. The huge recliner was piled high with two extra pillows, as if they knew all about my back troubles. So I sat. Bryce took his seat at his desk and sat facing straight ahead, or north, toward the hunt; toward the buffalo. Demetria had simply disappeared like a prairie sprite.

Before Bryce could even utter a word, I anxiously asked the $64,000 question: "Did you harvest yet?" It was late morning at the end of the first week of July. I was supposed to have arrived nearly a week earlier, but my daughter's high school graduation and another commitment had held me up. Since harvest was my sole mission, I was particularly anxious about arriving in time to participate. I had already missed out on a wheat harvest the week before, and this was possibly my last chance.* But luck was

---

* Originally, I was going to help harvest wheat at the world-renowned Land Institute in Salina, Kansas, Wes Jackson's farm, where they have been studying sustainable methods of agriculture since 1978. But I was too late; the crop had been harvested. I was particularly interested in finding out more about Wes Jackson and the game-changing perennial grain crop he has been experimenting with for going on thirty years. The idea of perennial wheat, for example, is that it would not need to be replanted each new season, which in turn would help the soil, leaving it more natural and less impacted by heavy equipment and destructive tilling. For more about Wes Jackson, the Land Institute, and the perennial wheat crop, read his *Consulting the Genius of the Place An Ecological Approach to a New Agriculture* (Berkeley, CA: Counterpoint, 2010).

on my side, as rain and humidity had kept Bryce from cutting his heirloom Turkey Red.

"Not yet." He shook his head with a wry grimace and a fisherman's sense of fatalism. "We've been taking moisture readings all week, and we're still three points off. I like to harvest at twelve. It was fifteen this morning."

"I saw the field across the street; it looked like the wheat had been cut," I responded, letting out an audible sigh of relief.

"Nagh, that's not my wheat. That's the neighbor's. He's a chemical farmer. That was ready a week ago. He just cuts the heads off."

And no sooner had all of the air been released from my lungs than he began to talk. (Let me just say at this point that the stereotype of the laconic farmer who speaks only when spoken to, in utterances no larger than a single syllable, is a complete and total myth. Farmers, or at least the ones I met, are as long-winded as the day is long.)* Although there were neither traditional Cheyenne drums beating in the distance or a *Heya Heya Hoa* chorus to make me feel like I was inside a sweat lodge participating in some indigenous harvest ritual, the quiver of Indian arrows, the big black felt Stetson with the feather sticking out, and the excessive heat created an undeniably native ambiance. But more than anything else, it was the cadence of Bryce's voice, his hand gestures, and the stories he told that slowly oriented me. In the sweltering midday heat, without air-conditioning, I sat in my well-cushioned chair listening in testicular-crimped humility as he told me the stories

* Lisa Hamilton offers an excellent example of the stereotypical loquacious farmer in her book *Deeply Rooted* (Berkeley, CA: Counterpoint, 2009).

of his life and his people and his land in a lilting and compelling voice.

At first I tried to record the conversation on my little Sony tape recorder, but I quickly realized that there wasn't a recording device on the planet big enough to capture all that he had to say, so I turned it off and put it away for good. For the remainder of the morning and a good part of the afternoon, he talked, facing the north wall, nary glancing in my direction. He told me about the land and about the people and about the wheat . . .

The roughly one-thousand-acre Stephens Farm lies between Beaver Creek to the north and the North Fork of the Solomon River to the south. Beaver Creek and the two-pronged Solomon River are part of a formation of western-flowing rivers called the Ladder of Rivers, which were formed millions of years ago after the Laramide Revolution, the geologic process that created the Rocky Mountains. Like the fully splayed hand of a bass player, this nexus of waterways reaches from southern Oklahoma and the panhandle of Texas clear up to western Nebraska. The latticework of rivers are all within a day's travel of one another, which afforded the Cheyenne and other tribes of the region who often traveled up and down the High Plains convenient water access.* From the front yard of the farm looking south, the eye is naturally drawn across the rolling wheat fields toward the gray-green ridgeline of the south side of the North Folk—no doubt a comforting sight to any horse-backed traveler looking for shade or sustenance at the end of a hard day's ride.

---

* To really get a sense of this land and the rivers, read Harry E. Chrisman's *The Ladder of Rivers: The Story of I. P. (Print) Olive* (Denver: Sage Books, 1962).

Bryce Stephens was one of ten children—seven brothers and two sisters—born in a Roman Catholic family of German lineage. I learned all about how Bryce's maternal great-grandparents came to Decatur County after the Homestead Act of 1862 was passed and how several of their offspring farmed in the area. I learned that Bryce's dad, Andrew Stephens, and his mother, Rosalia Stephens, bought their own land more than fifty years ago by getting a loan from the bank. They paid off the loan by selling wool, milking cows, growing wheat, and raising ten children who worked for free. "That is the way of most farmers out here in the open. They loved their way of life, loved God, and loved each other." After his father and mother passed away, the children inherited the land, but most of the clan didn't want to farm it, so Bryce and his brothers had to buy out the other siblings' shares. Today he alone farms it, growing organic wheat and alfalfa and raising chickens, cows, and bison. His journey back home to the farm is more than just an interesting story.

Returning home from the Vietnam War, disillusioned and disgusted by his government's policy in Indochina, he joined the antiwar movement. He was there in April 1971, along with John Kerry and other Vietnam Veterans Against the War who went to Washington to demonstrate and speak out against the senseless death and violence going on in Southeast Asia. He heard Kerry speak at the Capitol building. Bryce's antiwar activism eventually led him in 1973 to Wounded Knee, on the Pine Ridge Reservation in South Dakota, and smack-dab into the heart of the American Indian Movement. Traveling with fellow Vietnam vets, Indians all except for himself, they drove from Denver to Pine Ridge to offer their support to the armed resistance movement that was taking place at the infamous massacre site. The events that unfolded there (he didn't want to talk too much about it other than to say he was thrown in jail by the FBI and prosecuted, harassed, and discriminated against by his own government) ultimately provided him with his totem (buffalo) and direction (organic farming) in life. The Lakota had accepted him as one of their own. Indigenous culture was at the core of the Stephens' farm.

As for Turkey Red wheat, it is an heirloom crop that was brought over to the United States from Russia by the Mennonites. The story goes that Catherine the Great helped the Mennonites of Europe immigrate to Russia to avoid military service, and in exchange they were to raise their Turkey Red wheat for her countrymen. By 1870 the Mennonites no longer garnered any privileges for their farming practices and decided to immigrate once again, this time to Kansas. To avoid having their precious wheat confiscated at the border, the women sewed the seeds into their undergarments. The direct descendants of those seeds are the same as those reaped and sown by Bryce. Turkey Red wheat is a winter wheat, which means it's planted in the fall and needs to overwinter, in other words, freeze and go dormant in order to germinate. When the snow melts, little heads of wheat are there under the snow, just waiting for the sun to make them grow tall. The wheat is highly prized by bread bakers for its protein-rich qualities.

At some point during our talk, Demetria appeared with a plate of food and set it down before me on a folding tray. It was a welcome sight, as I had eaten nothing all day but a stale Danish. Hungrily shoveling the food in, I listened as the lessons continued . . .

I learned all about biodynamic agriculture, a practice that Bryce stringently adheres to on his farm, which has caused him no end of trouble. Surrounded by "chemical farmers" and "resource extractors," as he refers to them, he alone practices biodynamics, which he believes is the most sustainable and "organic" system of farming there is today. Biodynamics is a method of organic farming founded by Rudolf Steiner, also the creator of the Waldorf schools.* Biodynamic farmers regard and treat their land like an individual organism, which they believe is endowed with "archetypal rhythms." They scrupulously avoid the use of artificial fertilizers and toxic pesticides and herbicides that have systematically destroyed our country's soil fertility. For example, to fertilize his fields, Bryce uses the BD 500 method. He gathers loose, green bison droppings from lactating females, which droppings are loaded with a certain strain of bacteria, and packs them into hollowed-out buffalo horns. He then buries the horns during the cold months, between the winter solstice and the spring equinox, into a pit eighteen inches deep with the horns all pointing east. When it is ripe, it is removed, mixed with water, and sprayed onto the fields before the wheat is planted, at a ratio of about one-quarter cup per acre.

---

* The philosophy of the Waldorf education is to help young people become freethinkers who are morally and culturally responsible; moreover, it seeks to help them fulfill their unique destinies through the cultivation of a form of thinking independent of sensory experience that includes imagination, inspiration, and intuition. To learn more about Rudolf Steiner, biodynamics, or Waldorf education, read Steiner's *Agriculture Course: The Birth of the Biodynamic Method* (Forest Row, UK: Rudolf Steiner Press, 2004), and *The Education of the Child: And Early Lectures on Education*, Foundations of Waldorf Education, 25 (Hudson, NY: Anthroposophic Press, 1996).

It was at that point that Bryce revealed that he and fourteen seed businesses, thirty-two farms and farmers, and thirty-six agricultural associations, for a total of eighty-three plaintiffs, had filed a lawsuit against Monsanto, the world's largest seed company. He pulled out the original document filed in New York City by his attorneys, a nonprofit legal group representing all the plaintiffs in the case and calling itself the Public Patent Foundation (PUBPAT). It was all neatly bound in a three-ring binder. I flipped open the book and read the first page:

INTRODUCTION

1. Society stands on the precipice of forever being bound to transgenic agriculture and transgenic food. Coexistence between transgenic seed and organic seed is impossible because transgenic seed contaminates and eventually overcomes organic seed. History has already shown this, as soon after transgenic seed for canola was introduced, organic canola became virtually extinct as a result of transgenic seed contamination. Organic corn, soybean, cotton, sugar beet and alfalfa now face the same fate, as transgenic seed has been released for each of those crops, too. And transgenic seed is being developed for many other crops, thus putting the future of all food, and indeed all agriculture, at stake.

2. Plaintiffs in this matter represent farmers and seed businesses who do not want to use or sell transgenic seed. Plaintiffs are largely organic farmers and organic seed businesses, but also include nonorganic farmers who nonetheless wish to farm without transgenic seed. Plaintiffs are increasingly being threatened by transgenic seed contamination despite

> using their best efforts to avoid it. This causes Plaintiffs to fear that if they do indeed become contaminated by transgenic seed, which may very well be inevitable given the proliferation of transgenic seed today, they could quite perversely also be accused of patent infringement by the company responsible for the transgenic seed that contaminates them. Thus, Plaintiffs bring this action to protect themselves from ever being accused of infringing patents on transgenic seed.*

I stopped reading and shook my head, *disgustipated*: "Patent infringement? I don't get it."

"They're suing our farmers for patent violations because pollen from their genetically modified crops, growing in nearby fields, blew over the fence and contaminated our seeds. And once those genes are in our seeds, they lay claim to them."

"Is this something that's happening to your Turkey Red wheat?"

"Luckily our wheat has a lot of variation in its genetic structure, but we're trying to preempt them. They're already in ninety-five percent of soybean and eighty percent of corn crops grown in the U.S. They certainly know that whoever controls the seeds controls the world!"

I discovered yet another injustice. There is a full-scale, round-the-clock conspiracy against the obvious, and it is well financed, professionally managed, and magnificently effective. So, in addition to evolution deniers, global-warming pooh-poohers, omniscient-free-market fundamentalists, mountaintop-removal

* To read the entire lawsuit, go to http://www.pubpat.org/assets/files/seed/OSGATA-v-Monsanto-Complaint.pdf.

mountebanks, atomic-energy-is-safe beguilers, clean-coal con men, corporations-are-people-too lobbyists, prisons-keep-us-safe fearmongers, cell-phones-don't-cause-brain-tumors liars, one-hundred-billion-plastic-bags-a-year-is-not-a-problem ignoramuses, guns-don't-kill-people-people-kill-people misanthropes, a-college-education-is-worth-the-exorbitant-price-you-pay sophists, American-health-care-is-the-best-in-the-world extortionists, the-history-of-the-English-speaking-peoples-is-beyond-reproach magniloquents . . . I now had GMO-seed-is-the-only-way-to-feed-the-world megalomaniacs to add to the list.

Then without comment he was outside in the 100-plus-degree heat, under the baking-hot sun, reinstalling a truck radiator that he had recently repaired with his own welding equipment. "Brazing," as he called it. I stood nearby as he slid the radiator into place, the quarter-sized brass splotches glistening at me as if to say, "All fixed." Demetria was underneath the truck replacing a broken starter motor. It was a heartwarming scene, I tell you, the Prairie Deity's grease-stained boots and pant legs sticking out from under the chassis, Bryce standing on the bumper, bent inside the hood of the truck as if it were about to bite him in half, eight dogs lying in whatever odd slice of shade they could find, numerous geese squawking the squawk and waddling the waddle and sticking their long necks in where they weren't needed, chickens bawk-bawking and an occasional rooster cockadoodle-dooing, cows chewing the cud and watching disinterestedly from over the fence. And I? I just stood there with my hands in my pockets, of no use to anyone, least of all myself. I didn't even have enough sense to get out of the sun.

So I decided to find the buffalo that Bryce had mentioned he raised somewhere on the farm. Knowing that there was a thousand acres of Stephens land out there, I was about to ask, *Say, Bryce, where do you keep your buffalo, anyway?* but the thought of disturbing him with such an asinine question stayed my tongue. Instead I put my head down and walked north, through the car- and truck- and engine-debris field, away from the house and the chickens and dogs and geese and cows and shade, toward a thousand acres full of wheat and sun and glistening Kansas, glorious Kansas.

For nigh on an hour I sauntered along the wire-fence lines that separated the pastures from the cow paths, the wheat from the alfalfa, the long from the short, the me from the them. I was struck not only by the sublime beauty of the landscape, such as the statuesque nobility of the whiskered wheat heads tickling the endless blue day or the rutted swales lush with teal-colored grass and full of six-foot-tall, perpendicular sedimentary ledges, or the tawny long-eared hares that burst out of the ground like improbable fairy-tale elves, or even the thrilling black and slanted profiles of a small herd of grazing bison on the far edge of the horizon, but also, more than anything else, by the never-ending assortment of rusted-out and decommissioned farm machinery left to rot out in the elements. I counted three old combines, like Rube Goldberg monstrosities, all with giant windshields, ten-yard-long headers, and man-tall tires. The contraptions ran the whole gamut of eras. And in between them there was a yellow school bus, numerous planting or cultivating machines, and other odd farm vehicles and equipment far beyond my capacity to identify, all lying out in the fields, furry with rust and grass growing up over their hoods. The sight of these old farm vehicles left out to pasture made me understand that farming is an arms race. And as I stood there ruminating on this bitter thought, just inches from the electrified fence (the same one that, moments later, would add an additional one hundred volts of pure coal-fired electric energy to my sweet and sour epiphany), looking at the small herd of bison a quarter mile to the north, I recalled a phrase that Bryce had used several times during our discussion: "seeing the elephant." He defined the phrase as one used by the pioneers to describe their utter excitement and abject disappointment as they moved west across the land, likening their adventures and misadventures to "seeing the elephant."* "Seeing the elephant," *Yes! The perfect metaphor for life down on a Kansas farm*—"ZZZZZAP!"

Slightly light-headed, more from the extreme heat and sun than the electrification, I headed back to the work zone. I came upon the two of them gathered head-to-head around the starter motor. Bryce had it clamped to a heavy-duty vise on an old outdoor workbench, and he was pulling on the end of it with Vise-Grips, trying to realign it or reposition it, but without much luck. The harder he pulled, the more pissed off he waxed, until—"*Snap!*" That was it. Time to quit working on the vehicles.

"Let's take a moisture reading."

* It is believed that the origin of this phrase comes from an old tale about "a farmer who, upon hearing that a circus had come to town, excitedly set out in his wagon. Along the way he met up with the circus parade, led by an elephant, which so terrified his horses that they bolted and pitched the wagon over on its side, scattering vegetables and eggs across the roadway. 'I don't give a hang,' exulted the jubilant farmer as he picked himself up. 'I have seen the elephant.'" See http://www.goldrush.com/~joann/elephant.htm.

Next thing I knew, all three of us were out in the wheat field a mile from the house. I was holding an old scythe that Bryce had sharpened, and it was my job to cut down a bucketful of wheat heads. The antediluvian contraption, made of wood, curved and sleek and two-handled, looked like it would be easy to wield, but it wasn't. I just couldn't find the right swing style. It was awkward and asymmetrical and not at all like a baseball bat or a golf club. Normally, he explained, the combine would cut the wheat, remove the hulls, and separate the chaff, but we were just there to take a moisture sample of the kernels and needed only a cup full of seeds for an accurate sample reading. As I fecklessly sliced into the hip-high stalks with the scythe, he talked about the old days and how they'd cut the wheat just the way we were doing it. (They must have been a lopsided, bent-to-the-left-or-right people!) But then along came the horses with the threshing machines behind, and now the combines, growing in size and capacity year after year, which had me replaying my realization an hour before about farming being an arms race and the old rusted machines like "seeing the elephant."

Under the shade of a little canopy set up alongside the farmhouse, with a picnic table in the middle, we proceeded to remove the hulls and "separate the wheat from the chaff." It was a particularly slow process in which you took the tiny heads and pulled the seeds off with your fingers, whiskers and all; then you rolled them in your palms, one against the other, until the hulls wore off. Next, you put them in a round, handwoven straw basket and literally blew off the chaff. The end result was a morselous torpedo-shaped kernel, golden brown but with a slight ruddiness to it, about twice the size of a sunflower seed. Bryce kept biting into them and shaking his head: "Not ready. Not ready."

The moisture reader was a fairly sophisticated-looking contraption—as big as a bread box, it was perfectly square with a handle and a meter and dials and a thin little slot in the top where the seeds went in. When we had a cupful of seeds ready, Bryce measured them out and then impatiently dumped them down the slot. Demetria quickly set the meter and then both of them sat back for a moment and waited.

"Fifteen," Demetria called out.

"Too moist. I don't cut till it's twelve and a half."

Back inside the house we went. I assumed my seat in the pillowed chair, and Bryce resumed his seat at his desk and took to storytelling again, giving me more details about the Wounded Knee affair and his various peregrinations throughout the United States. Meanwhile, the heat refused to abate even half a degree. Someone had placed a tall oscillating fan in the middle of the room, and I was literally counting the seconds between the revivifying microbursts to my face. Every ten minutes I got up and went into the kitchen to get a drink of water, walking and listening and uttering "Uh huhs," between desperate and greedy gulps. Somewhere along the line another huge heaping plate of food was set before me, and I ate perfunctorily as the sweat lodge ceremony continued unrelenting, my heart taking the place of the big drum.

Bryce talked at length about the water: the taste and quality and whence it came. I noted that there was a complex mix of flavors to it, some sweet tasting, but some a little rotten-egg-flavored as well. "That's the sulfur," he explained. He lectured me about the Ogallala Aquifer, which underlies much of the High Plains, from Nebraska to Texas. Without the Ogallala

Aquifer, large-scale farming in the High Plains would be impossible because the amount of precipitation is not sufficient to grow water-intensive crops like wheat and corn and soybeans and cotton. However, areas of the High Plains are increasingly draining water from the aquifer at a rate too fast for it to recharge itself. Eastern Nebraska, southern Kansas, and northwestern Texas are particularly negligent in this regard. There is a great gnashing of teeth over this, he explained.

He talked about playas, dry lake beds full of salt and minerals that form after evaporation but are devoid of vegetation. He had pointed one out to me on our ride over to cut the wheat, but the sight of a neighbor's stray cow interrupted him, so I never got the complete story. Anyway, these playas form when depressions in the landscape get filled up with rainwater after a storm, creating ephemeral ponds or lakes. When all of the water evaporates, a dry lake bed is formed. Certain salt-tolerant plants grow around the circumference of them, and livestock can and will eat them in the wintertime. At that point Demetria joined the conversation by turning on the big flat-screen television and plugging her laptop into it in order to show me an up-close, bird's-eye view via Google Earth of one of the playas on their farm. I was amazed by how perfectly round these playas were. And as she zoomed in on one particular specimen, the lady of the house, Bryce's wife, Demetria's mother, Lin Davis-Stephens, arrived home. And before she took two steps inside the house, she had joined the conversation:

"Oh, you're teaching him about the playas! Did you tell him about the fairy shrimp?"

"The fairy shrimp?" I sneezed, inadvertently activating a neuron transmitter that flashed a photograph of a tuxedoed Truman

Capote at the Black and White Ball standing next to Katharine Graham, the latter a full head and a half taller.

Lin moved sideways into the center of the room, her plastic bags of groceries dangling from both her arms. She continued right on talking about the fairy shrimp: "After a rain, tiny little shrimp magically come to life in the playas after months or even years of lying dormant."

I sat scratching my head, trying to guess whether this was real, or another leprechaun tale.*

"So then if you go out there and pour a bucket of water on the spot, will they come to life?"

"Hmmm? I've never tried that."† She stood there for the longest time, her whole face lit up by the miraculous thought of the fairy shrimp; perhaps she was imagining herself pouring a bucket of water on the playas and seeing the shrimp come to life. Eventually she snapped out of her reverie and dropped her bags as I climbed out of my chair, and we both introduced ourselves. Silver-haired, rosy-cheeked, wearing lots of Native American baubles and beads, she reminded me of the singer-songwriter Joan Baez. I noticed immediately that her sophisticated manner of speaking harked back to East Coast hippie communes rather than High Plains Kansas. But she was a Kansan, born and raised

---

* On my one and only trip to Ireland, after a night of wild debauchery my wife and I stood out under the stars in the middle of a deserted street talking with an animated, drunken Irishman we'd met and who insisted that the soft tapping sound that we *couldn't* hear off in the distance was, in fact, a little green man, cobbling shoes . . . "You do hear it, yes you do, you do indeed . . . He's a-cobblin' the shoes, don't you know . . ."

† For more about fairy shrimp, here's a website that explains it all: http://faultline.org/site/item/the_playa_isnt_lifeless/.

in Wichita. A polymath, Lin Davis-Stephens is an anthropologist, a bar-certified lawyer, a linguist, a geographer, a teacher, a mother, and, as I would find out in the coming days, an honest-to-goodness, real-life pioneer cook woman.

She had Mexican food in the bags, and she proceeded to divvy it up on plates and pass it around. Even though I had just polished off a huge meal, I ate two enchiladas and a mound of chips and salsa. In the Stephens home, food, like water and air, was *conditio sine qua non*.

That night in the stifling heat, up in the second-floor bedroom without even a suggestion of a breeze entering my small bedroom through the wide-open window, I tossed and turned, unable to sleep. At one point I got off the bed and knelt with my face against the screen window, looking out at the night, beseeching coolness. Suddenly, an amazing event—a shooting star streaked low across the wheat field, but an instant later, it ricocheted back 120 degrees and then turned another 90 degrees and disappeared. *What the hell was that?* It took my breath away and caused the hair on the back of my neck to stand on end. I stood up, trying to determine if I was dreaming. I was definitely awake. I calmed my thoughts, knelt back down, and continued to look out at the field for an answer. Then it happened again, only farther off in the distance. *Was this, at last, the War of the Worlds?* I remember thinking to myself. Then it happened right in front of my nose! *Ooooh, lightning bugs!* Either High Plains lightning bugs are far more dynamic, stay lit longer, and glow far brighter than the little wimpy blinky fireflies we have back east, or that optical/psychological illusion effect that I had described earlier regarding the brain creating larger-than-life reality really kicks in when the lightning bugs dance across the night sky.

I was up early the next morning, around six or six thirty, but I was the last one out of bed. Bryce was at his desk, going over some documents. Demetria was nowhere in sight, no doubt engaged in a half-dozen projects outdoors. Lin was banging around in the kitchen. I noticed that over by *my* chair, sitting on top of the folding table, was a plate of food kept warm by another plate turned upside down on top of it. On the end table next to my chair sat a thermos of coffee and a cup. Who knows how long they had lain there. I took my cue and removed the top plate to reveal three farm-fresh eggs sunny side up, thick slices of organic bacon, thick slices of wheat toast, and a great mound of potatoes. It took some doing, but I cleaned the plate.

Lin joined me as I ate, and I learned how she and Bryce had met more than thirty-five years ago at the university during a meeting of the American Indian Student Association. They were both doing treaty research as law students, Lin on the Cheyenne and Bryce on the Lakota. Then Bryce joined the conversation, and the sweat lodge ceremony continued, hour after hour, with that same feeling in the back of my head of a distant drumming growing slowly louder and closer. We sat there until lunch, at which point we ate a sit-down meal in the dining room/living room. It was a feast fit for a king, with homemade bread and canned items from their pantry.

After lunch we went back into the wheat field to take another moisture reading.

"Fourteen and a half."

"Nope."

Bryce and Demetria began work on the thirty-foot-tall galvanized storage bins, preparing them for the wheat seeds. I helped as best I could, which was mostly holding the ladder while Demetria worked up on top. "We" filled cracks with black goop where ants were threatening to invade. "We" put a huge drying fan in place in the back of one of the bins, fashioning a screen over the opening so no animals could decapitate themselves. "We" set up a forty-foot tube from the back end of the tractor to the top of the bin that would shoot the seed up from the pile on the ground and then down into the storage unit. The propellant was an auger, attached to a revolving rod on the backside of the tractor, directly below the driver's seat and above the back axle. This revolving rod, called the power takeoff, is the modern-day workhorse that perpetually spins and moves all manner of material. It was during all that work that I had another revelation: *farmers are engineers.*

Bryce seemed to have an uncanny ability to eyeball problems, devise solutions, and work them out in process. For example, he needed to make a sleeve for the top of the chute that would introduce the seed from the tube top down into the bin. He had Demetria take some measurements, and then they went back to the vehicle-debris field, where every tool and miscellaneous part ever invented by mankind was somewhere within reach, rusted but no less serviceable. He found all the parts he needed—chains, galvanized ducts, wingnuts, bolts, steel rods, etc., and with his welding equipment he put it together just like he'd made one a hundred times before. Amazing.

By the time the sun was setting, a dark mass off to the southwest had filled up the horizon line completely. Lightning far off in the distance was ramifying across the welkin, up and down and sideways, getting closer and closer. We ate dinner. Once again, what a spread! There was bison (from their own herd) meat loaf, mounds of potatoes and vegetables (from their garden), and mulberry (from their trees) pie with fresh ice cream (from their cows) for dessert. As we ate, I got to thinking about the whole foodie movement, and all the fanfare about fresh herbs and spices and heirloom vegetables and grass-fed beef and free-range chickens, and how overstated it all must seem to the Stephenses, who grow and raise and produce and cook and eat their own organic food every day. A meal at their table is as extravagant and sumptuous and *sublime* as any saffron-and-garlic-infused entrée served at Chez Panisse.

To watch Bryce eat a meal is worth an admission fee. Head down, focused like a monk in meditation, he *harvests* his meal. The food moves from plate to mouth in one fluid motion, with occasional rest stops to refill the eating surface, pour more water, or nod in satisfaction. Both hands are involved in the process. There's no more glorious appetite on this planet than that of the person who eats the food he grows and kills and catches.

After dinner, we watched the Weather Channel on the large-screen television. The scrolling text at the bottom of the screen was reporting that a tornado warning was in effect for our county. Lin, who was following the Doppler radar, turned her laptop screen around to show us the dark red cell formation that we were reading about. It was about thirty miles to our southwest and so red it was black . . . and on a trajectory straight toward us.

I stared at the screen in petrified disbelief. There was a long, heavy silence. All eyes were on me. Finally Bryce spoke: "Did you know that the Clutter family, you know, the one that was murdered and written about in the book *In Cold Blood*, was from Holcomb, just south of here?"

"Huh?" The comment distracted me from my simmering hysteria only for a moment. Quietly terrified, I bolted to my feet. "So what do we do now?" I asked, trying to sound cool but coming across as a flight risk.

"We go and watch for it," Lin replied calmly, smiling from ear to ear.

She stood up and I followed her outside. Demetria had more important things to do than watch for tornados, like make more ice cream, so she remained inside. Bryce was preparing a tobacco pipe to smoke while he read through a magazine.

As Lin leaned back against the trunk of one of the cars out in the yard and looked toward the foreboding sky, I suddenly realized what a sadistic twist of fate this all was. As a child I had suffered from terrible nightmares as a result of the tornado scene in *The Wizard of Oz*. For years that monstrous tornado was the featured bad guy in many a midnight horror ride. Later on, somewhere in my tweens, after I'd read *In Cold Blood*, I had terrible nightmares of another sort—murderers! So here I was, forty years later, out on the High Plains of Kansas cavalierly scanning the sky for tornadoes and the dirt road for headlight beams heading straight toward me.

"So, I take it you guys have a storm shelter?" I asked in a hopeful tone.

"No."

"No! So what do you do if a tornado is coming?"

"Well, we have a little root cellar under the pantry. We go down there and wait it out and pray. Bryce doesn't like to be confined, so he just sits up in the house." She had turned around and was looking at me with a warm and comforting smile. I was really starting to feel nauseous. She then offered me this tidbit as solace: "Don't worry. Richard, if it is tracking our way, the sheriff will call to warn us."

At that moment, perhaps sensing my mounting hysteria, one of the dogs came over and nuzzled my hand with its nose. I knelt down and began petting it frantically, thinking almost out loud: *Kansas . . .*

The countryside was hotter and moister the morning after the storm. There would be no harvest today, either. The land oozed, and the distant ridge of the North Fork of the Solomon River had turned dull and gray. I sat in my chair and listened to more stories, writing as fast as I was able. I learned about weeds and how weed science is a hot topic in agriculture. Bryce confided that he thinks he has found a weed crop that will outcompete all other weed crops, and that will work as weed inhibitors–cum–crop enhancers, thus making him a very rich man. He wouldn't show me the weed, however.

After lunch I learned about the Cheyenne sun dance ceremony and about the sacred pipe and fasting. I thought I had read that during these ceremonies warriors stick bones through their skin and hang from leather straps, and I asked him about that, but he vigorously shook his head, assuring me that the Cheyenne don't practice this type of self-mutilation. "They are a land culture," he said. "It's important that they keep their feet on the ground." The words "land culture" shook something loose inside. Native culture is indeed a land culture, one that goes back to our species' beginnings. And farming, too, is a land culture, though not as old. For a hundred thousand years humans lived a land culture. Now, in the blink of an eye, we have turned away from the land, turned our back on that culture, and assumed our seats behind the wheel on inflated tires or on the wing. Economics, not land, is the culture du jour. Wendell Berry put it best: "It is becoming harder to remember—especially, it seems, for most economists—that our lives depend upon the economies of land use, and that the land-using economies depend, in turn, on the ecosphere." But the further away society gets from land culture, the more precarious our existence becomes. Thank God for the Cheyenne and the Stephenses, who still know and still practice land culture, the true culture of our species.*

Later in the afternoon Bryce and Demetria and I took a ride around the farm, eventually locating and then pulling up alongside the small herd of buffalo. As we bumped through the pastures of blue-green grass and cacti and exotic flowering weeds, four of the dogs trotted along behind the van. It was reminiscent, in a way, of a cattle drive, only instead of being on horseback we were comfortably seated in an air-conditioned van. We were parked there with the windows open, the dogs waiting for Bryce's command as we spoke to the buffalo. Sitting there, look-

* I later discovered that Lin had written a very similar thought in the paper she presented at an anthropology conference in China (cited earlier), to wit, "Indigenous peoples maintain a relationship to living beings in their environment with commitment and obligations of dynamic consequences throughout generations carrying the traditional knowledge systems. Now is the generation to make a conscious effort as individuals, local communities, and in the dominant society to better understand and live within the dynamic carrying capacities of local ecosystems. Indigenous knowledge needs ancestral/ecological wisdom to remember how and when and what is suitable. Ecological wisdom is the practice of indigenous knowledge."

ing with quiet admiration at his herd, Bryce was as proud an animal husband as I had ever witnessed. The herd was a tattered-looking bunch, as their heavy fur in the hot weather was falling off in batches. One of the females was pregnant, and he pointed out her swollen sides to me. I could sense what Bryce and people of his culture feel today at the sight of buffalo living out on the High Plains: it's a spiritual calling, with a sort of healing power as well. Buffalo alive and well on the High Plains adds another inch of arc to Black Elk's broken hoop.* Bryce suddenly gave a loud whistle, and the dogs exploded in barks and began running around the buffalo herd, causing them to stampede. It was an awesome sight, the buffalo at full gallop, all legs and heads, running as if . . . as if . . .

The native drums were beating louder and closer than ever as I walked down the stairs the next morning and entered the sweat lodge. My plate of food was in its regular spot, but one thing I noticed was out of place: the barometer/thermometer. It had been removed from the wall and placed in the middle of Bryce's desk. My eyes zoomed in on the barometer, which read zero, then they shifted to the thermometer, which stood poised at 90 degrees. Walking outside nearly took my breath away. It would be as hot a day as I had ever lived through.

* The Lakota medicine man Black Elk used the symbol of the broken hoop as a metaphor to describe the suffering that the native people had endured as a result of the invasion of white culture. But he had a vision that native people would begin to heal. The closing of that Sacred Hoop is the symbol of that time of healing.

At two o'clock, the temperature hit a state record, 111 degrees. We took a moisture reading soon afterward.

"Thirteen and a half," Demetria read aloud.

"Let's cut. We can blow-dry it down once we get it in the bin."

Then we were up on the combines, seven feet off the ground. I was standing on the metal shelf, like a fire escape, outside the single-passenger compartment, holding on to the railing and looking down at Demetria, who sat at the controls like a jet pilot, revving the engine and checking the functionality of her dials and levers. The motors of the two beasts were as loud as locomotive engines. Bryce was thirty feet to our port side, standing on the ground and adjusting something on the other combine. When I looked back a few moments later to see what he was doing, he was right below me, seated at the wheel of a pickup truck, shouting and pointing at me through the half-rolled-down passenger-side window. I couldn't hear what he was saying and so turned to Demetria for a translation. She jumped out of her seat and stood next to me on the shelf, crouching down and shouting back. They exchanged words, none of which I could hear. She then stood up, turned to me, her face not but three inches from mine, and said above the roar of the engines with a little twisted smile, "He wants you to drive."

"*Me!*"

The next thing I knew, I was seated behind the wheel of that magnificent beast, driving it down the road like I was heading to war, following in the wake of the other magnificent beast in front of me, giant wheels spraying dust clouds and obscuring my vision, the huge header in front of me hanging like a rolled-up tongue as wide as the two-lane road itself and threatening to lead me astray at any moment. There was a lot of play in the steering column, and I had to use all my powers of concentration to make sure the combine didn't bounce and swerve out of control as I trundled and bobbled along the road at twenty miles per hour, bringing up the rear of a harvest armada. And in my head was only one set of instructions: "Whatever you do, don't step on the brakes!"

"Yee-haw, Kansas! Fucking Kansas!" I yelled at the windshield.

And then I was out in the wheat field, right behind her . . . and the headers were down on the crop, and the teeth were cutting and thrashing, the rollers churning and casting the wheat heads back up into the belly of the beast, as the sky overhead glowed blue and the sun on top raged fire and the engines out front roared power and the sea of wheat down below sang siren songs and the big drums were drumming in my head so loud that I couldn't tell if I was driving or not. And on we rolled across the acres like mechanical waves upon the earth, round and round and up and down, harvesting wheat. Wheat! In Kansas!

Somewhere along the way the equipment malfunctioned, and Bryce, who hadn't drunk a drop of water the whole way through, looked a little weary. We unloaded our harvest into the back of a large dump truck near the storage bins. I watched as a tube that lay flush against the body of Demetria's combine suddenly came to life. With a push of a button, the galvanized tube, a foot in diameter and about eight feet long, robotically extended slowly out over the bed of the dump truck, and in an instant seed came flowing out. Then it was our turn. It was an arresting sight to see the transformation from gangly stalks of wheat to ripe red seeds ready for milling. (They talk about the gold standard. It should be the wheat standard.)

I had never felt so hot in my entire life. Yet I had never felt such intense pleasure at a task. We stumbled out of the machines and into the cool darkness of the home. I went to the kitchen sink and downed three glasses of water while Bryce collapsed into his chair. He was like a great desert camel, needing shade and rest more than water. He had really pushed himself, probably more than he should have, not only because he was over sixty but also because he suffered from a mutation called thrombophilia.* We sat for a while in quiet contentment, no one talking. Lin was there making sure we were all properly hydrated.

But there was no time to rest on our laurels; we hadn't finished the harvest. Within twenty minutes, we were all back outside working on the combines. Demetria was adjusting a feature she called the concave, while I had my head inside the belly of the monster, cleaning out wheat stalks that had gummed up the works. I had to pull my shirt over my face so that I could breathe; the dust inside was as thick as dirt. Soon I became aware of a caterwaul coming from the other side of the vehicle, and I stopped what I was doing to investigate. Apparently the wrongly set concave had caused a chain to break, and Bryce was barking up a storm, yelling fix-it solutions at Demetria.

Meanwhile, storm clouds were gathering . . .

Within the hour we were headed back to the fields to finish the job. I stood on the fire escape ready for another round of harvest, but a crack of lightning overhead dissuaded us, and we did an about-face and headed back toward the dump truck to cover up the seeds before the rains came. Just as we got there, the

---

* Thrombophilia is an abnormality that causes blood coagulation, which increases the risk of blood clots, mostly in the legs.

sky opened up, with cloud-to-ground lightning electrifying the moment.

Demetria and I jumped from the combines like swimmers from starting blocks. With two quick strides she was up and onto the side of that truck, pulling a plastic tarp over the seeds with celerity. I raced around to the other side to do the same, but the little ledge was too small for my fat feet. I tumbled to the ground. *No excuses for failure!* I immediately hopped back up and jumped onto the ledge and fell again. Utterly determined, I leapt again, this time gaining a toehold. Like a ballerina en pointe, but wearing work boots, I managed to tiptoe down the line as I pulled that tarp over the seed before she had to do it for me. Another crash of lightning had me jumping from the truck and hunkering down. Meanwhile the Prairie Deity was doing just the opposite—she had scampered up the ladder to the top of the stainless steel bin in order to secure the cover that was in danger of being blown open by the manic gusts of wind. With the rain coming down in sheets, and lightning threatening to vaporize us, the Prairie Deity was there on top of the steel monolith, forty feet in the air, defying Zeus's wrath. Lid secure, she climbed down while I held the ladder. Bryce, like a general, looked on with satisfaction from his seat inside the combine. And all around the booming of the big drum in the sky was punctuated again and again by jagged exclamation marks of electricity . . .

My time at the Stephens Farm had come to an end. The morning after harvest I was packed and ready to go. We were all going into town, and they were going to take me on a tour of the city of Colby, culminating with a visit to the Prairie Museum. But before we left the farm, Bryce took me around to the back of the house, back behind the chicken coop to a little cluster of mobile homes, three in a row. I hadn't really noticed them before. We went inside one that was filled with books. It was a large collection, varied in subjects from organic gardening to philosophy to literature. But he didn't take me there to show me the books; he wanted to talk about his children. There were four, one boy and three girls, Demetria being the youngest. He showed me pictures of them, and artifacts from their youths. The other three were all grown up now, living out of state, with families of their own. But Demetria was the ultimate topic of the conversation. He led me into the back part of the trailer and showed me her art studio, and all of her stunning paintings, some of them surreal. There were sculptures and photographs as well. It was an impressive accumulation of work for someone so young. One photograph in particular caught my eye: that of a single wheat head standing tall in the foreground against a background of a deep blue Kansas sky.

"Demetria is part of the deep map, and so vital to the life cycles here," he said. "She's going off to Lincoln to go to college," he concluded, more to himself than me, his tone wistful and subdued.

"Yes, I know" was all I could think to say.

Change was coming . . .

And as I headed west out of Kansas toward the winking sun, the drums had stopped beating, and inside the air-conditioned car my body felt enervated, different, like I had been sick and somewhere far away for a long time. Outside the *toxto* rolled on, climbing ever higher until somewhere around Limon, Colorado, when over a long and continuous rise, the Rocky Mountains appeared before me like a second-story continent carpeted in pure white. *God likes changes, too*, I understood, *and from the looks of it, the bigger the better.*

Six weeks after leaving the Stephens Farm, and after numerous unanswered inquiries about the fate of the wheat harvest, I finally received a reply from Bryce:

> Dear Richard,
>
> All of the wheat was harvested. It took until July 21 to finish the cutting. This year's harvest was very unusual because it took so long. The wheat is all in Bin #8. It was very hot and rained nearly every day, which made the grain high-moisture. We had to wait until the heads dried down to 12 moisture. We cut most of the wheat at 14 moisture and used aeration fans in the bin to dry the wheat. Since it was very hot, drying the grain was not a problem.
>
> I also harvested a neighbor's wheat about twelve miles away. He has a certified organic farm and has twelve acres. So one day I took the combine over there and cut it. The next day I went to get the combine, and on the way back a radiator hose blew. The combine's temperature gauge doesn't work, so the combine engine overheated, froze up, and stalled in the middle of the road. I took off walking, figuring I would walk home, about three miles. It was 109 degrees that day with no wind. The first farmhouse was

Mrs. Gillespie's, and she wasn't home. Three quarters of a mile away is Jeff Wahlmeier's place, so I set out for that. Nearly didn't make it. I staggered to the door, knocked, and collapsed from the heat.

Mary Wahlmeier eventually answered the door and immediately put water on me, wrapped a cold towel on my head, and gave me water. Jeff and his son Nick had shut down harvesting because of the heat and stood at the kitchen sink drinking water, nearly ran the well dry. We called Linda [he calls his wife by this name] and she buzzed over with the car that had air-conditioning and got me home and cooled off. But the combine was still in the middle of the road. So I went back because somebody might run into it in the middle of the night.

I was trying to get the engine to turn over and a thunderstorm came up—ninety-mile-an-hour winds, big hail, and three inches of rain in a half hour, so I drove home in that. Then the sun came out. Next day I got a tractor and pulled the combine home. My brother Paul, who is the custom combiner, with about nineteen combines, sent one of the machines—he has a John Deere—over to the farm, and with Demetria on the N5 Gleaner and Danny Stephens, my nephew, on the John Deere, they got all the harvesting done. I drove grain trucks. Can hardly sit in hot combines anymore. Got to get the air conditioners fixed for next harvest.

I am going to begin planting here in a few days, so some of the wheat will go back into the ground for next year's crop. The rest of the wheat is on inventory in Bin 8 for when Heartland Mill needs to grind some more flour for the Slow Food artisan bakers.

The PUBPAT lawsuit—There was a motion to dismiss by Monsanto. Pubpat responded with a motion that argued why the lawsuit should not be dismissed. And there was an amicus curiae brief—friend of the court filed by a bunch of organizations that are not on the plaintiff list telling the court reasons why the lawsuit should proceed. There is going to be a big demonstration on the East Coast calling for GMO labeling. And now I am involved in a Center for Food Safety lawsuit regarding Roundup Ready Alfalfa. There is an Heirloom Seed Expo and Food Biosafety conference going to happen pretty soon in California. I'll probably be planting wheat about then, but I would really like to go to that one.

Bryce

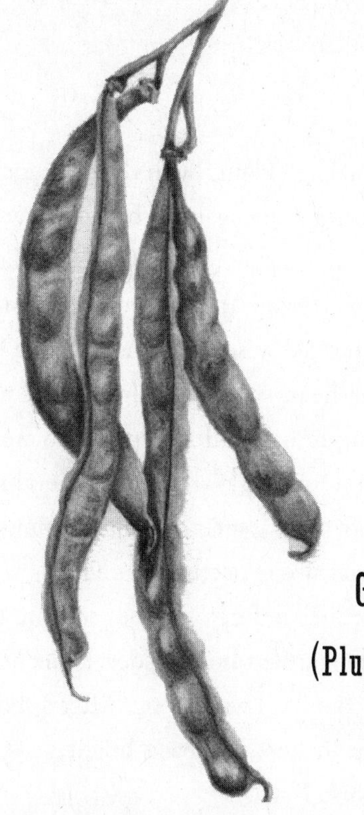

Green Beans
(Plus potatoes, squash, and arugula)

TANTRÉ FARM

CHELSEA, MICHIGAN

JULY 2011

A week after the High Plains, with the temperature nearly 100 degrees, I was on the road again.* This time I was heading west along the Ohio Turnpike, bound for Chelsea, Michigan. At Toledo, I turned north toward Ypsilanti, onto Michigan Route 23. At Ann Arbor, I headed west again on Interstate 94. Fifteen minutes later, I exited the interstate and drove three miles south before pulling onto a hard-packed and perfectly crowned and maintained dirt road. Driving slowly along, I found myself surrounded on all sides by rolling green farmland immured by tall hardwood forests. I was not expecting to find this landscape. I had assumed that the area a dozen miles or so from the Michigan Wolverines would manifest housing developments and strip malls, not lush farm country. I was wrong. After a sharp turn in the road and a steep incline, I spied a bright groovy sign up ahead that read: "Tantré Farm."

The white farmhouse was picture-postcard perfect, surrounded by shade trees on all sides. A tailless shepherd dog came out of the bushes to greet me, wagging its back end. Together we walked to the front steps. I knocked on the screen door, but no one responded. I called out. Nothing but chirping birds in the trees. Dog and I walked around to the back of the place, which afforded a partial view of the forty-acre farm.

---

* The oppressive heat persisted for the entire week.

Among the picnic tables and young trees in the backyard, there was a little gingerbread shed, ten feet by fifteen feet, two stories tall, built of wood and freshly stained a deep walnut brown. It had a large door reinforced with crossbeams, and it was wide open. I stuck my head in there, called out again, but found no one. To the right of the shed, fifty feet away, were two chicken coops, one on either side of a wide, wheel-rutted path. Directly behind the gingerbread shed and down the hill was a large bean field, its front and far side lined with long tables of seed-starter trays. The trays were laid out on plywood atop wooden horses, describing a leafy L. There was a young man watering them, and I was just about to trip down the incline and ask directions from him when a shoeless young woman in her early twenties, with short dark hair, dark complexion, and rosy cheeks, accosted me from behind. She wore a straw hat and big-toothed Chiclets smile.

"I'm looking for Richard or Deb," I informed her.

Without a word, she disappeared inside the house, reappearing a few moments later shaking her head. "I think Richard's down back," she finally spoke. And without as much as a "follow me," she proceeded to lead me along the path between the chicken coops, with her bright white teeth lighting the way. Her bare feet looked tough as shoe leather and about the same color. I had to ask:

"You always go barefoot?"

"Usually," she said, nodding.

"Does everyone else here?" I was thinking out loud, concerned that I might be obliged to go barefoot during my stay, because I'm a real tenderfoot.

"No, not everyone. Most people wear boots."

Past the chickens the path on our left was lined with flowers of all shapes, colors, and bouquets. Across from the flowers to our right were long rows of berries, the bushes as tall as we were. A few strides beyond that, we came to a small rise, and at that point the full extent of the forty-acre farm burst into view. What an absolutely gorgeous place! Dozens of acres of lush gardens bursting with multicolored abundance and glowing like manna out in the sweltering sun.

At the terminus of the path stood a red apartment/warehouse, with living quarters on the second floor. It had a huge, barn-style door that yawned wide open. We walked inside, where a half-dozen young people in their twenties worked diligently at tables, filling produce boxes with bright-colored vegetables. They seemed to be in a hurry as they hastened at their respective tasks. No one was talking, but everyone was smiling. No one paid me any mind, either, as I stood there watching, inadvertently scratching my head. Outside in the surrounding fields there were just as many people bent over and harvesting in the vegetable rows. I have to say, my curious reader, that it was a hotbed of tranquility around there.

"Anyone seen Richard?" the girl asked the group inside the work area.

Everyone was in the full flower of his or her youth, full of purpose and concentrating on the task at hand. I noticed, too, that they were all wearing straw hats. Something about that uniformity made me feel a little uneasy.

"I saw him getting ready to plant garlic with the tractor," another young woman finally offered without looking up, her hands moving with blurring speed.

"What does he look like?" I asked.

"Well, he's . . . he's . . ."

"Handsome. He's very handsome," another young woman holding an empty produce box said without much hesitation, walking up next to the first young woman, who was struggling to find the words.

(Okay, now at this point, perspicacious reader, I'm sure you're thinking the same thing that I was thinking at that moment—*Is this some sort of cult of personality? You've got the word* Tantré, *which is a variation on the word* tantra, *a Hindu or Buddhist word involving mantras and meditation, yoga and rituals, gurus and yogis . . . and all of these smiley young college-age people working like drones, straw hats, bare feet, shiny white teeth . . . describing the boss man as very handsome.* I will admit that I have a phobia of cults, having almost been kidnapped and reprogrammed by the Moonies back in college. I was feeling downright uneasy.)*

---

* During college, in Boston, I was going to meet a friend out in Cape Cod for the weekend, but I missed the bus. I decided to wait a couple of hours for the next one, so I went across the street to the Boston Common and lay in the grass, reading. A young woman came along, pretended to mistake me for an old acquaintance, and ended up inviting me to a party with free food. "Free food" are magic words to any college student, so five minutes later I was inside a lavish, million-dollar Beacon Street town house with twenty-foot ceilings, looking out over the Common, inhaling pizza and soda. There were at least three dozen people in the room, all talking in an animated sort of way. After pizza, we were invited into the next room, a huge parlor, for a little entertainment. For the next two hours, I was treated to a nonstop string of professional jugglers, jazz musicians, puppeteers, and proselytizers. (Yes, I'd already missed the second bus to the Cape.) This *carnival of fright* culminated with a lecture by "a professor from Cornell University," as they billed him, who with chalk and a blackboard proceeded to mathematically prove the existence of God. He ended his inspired speech with these ominous words: "And that's why I want you all to join me up in New Hampshire for the weekend to learn

I was about to ask, "Is . . . *he wearing a white robe*?" But I thought better of it and said instead: "Is . . . there any place I can get a drink of water around here?"

The girl who had led me there took me out of the warehouse and up to a little shed, where a water hose sat at the ready. I drank great gulps of the sulfur-flavored water, pouring some on my head and arms and gasping with cool delight, while remaining well on my guard against all things cultish.

Then I heard my escort say behind me as I played gurgle games with the water, "There he is!" I turned, and sure enough, not dressed in a white robe, but in the requisite straw hat, bare feet, jeans, and tattered shirt, was Richard Andres, handsome as a prince and walking toward me.

"So you made it?" he said, as if he had doubted me all along. He had a shy demeanor, but a waggish sort of wit.

---

more. A bus is waiting right outside to take us there now." That's when I knew I had been hoodwinked by a cult. I leapt to my feet, and as I did so, so did a half-dozen other people around me. I was surrounded. Only moments before I had thought that these people were just innocent partygoers like myself, but I could see in their eyes that they were all indoctrinates. They also looked like they were ready to jump on top of me, hog-tie me, and carry me out to their bus, kicking and screaming. I glanced around and realized that there were only a handful of "recruits" like myself in the room, and they, too, were completely surrounded by these folks who were trying desperately to lure them out to the bus. I spied the exit door thirty feet away, put my weight on the balls of my feet, gathered my fists in front of me, and uttered the following through gritted teeth: "Not going to New Hampshire. Leaving now, and nobody is going stop me." At that, the girl who had recruited me burst into huge crocodile tears, which enraged the group surrounding me even more, causing them to crowd in closer. So I pushed back through the crowd, creating a little seam for myself; then I darted, zigzagged, blocked, and burst out that exit door, knocking over a person or two and several folding chairs in my wake. The first thing I saw as I hit the sidewalk on Beacon Street was that idling bus.

"Of course, but I sure wasn't counting on this heat." We shook hands.

"Yeah, it's been pretty hot," he admitted, removing his straw hat and scratching his balding head with his dirt-caked fingers. Richard Andres, about five foot ten, forty-eight, has a round face, a lithe and well-balanced physique, shining gray eyes, and a winsome, at-peace smile—something like James Bond in farmer clothes. He then looked at me admiringly, sighed, smirked, put his hat back on, and looked at me some more. I understood the look: *Your turn to talk.*

"So where do I start?" was all I could think to say.

"How about in the potato patch?"

"Why not?"

I felt a little less on edge having met the man himself as I sauntered over to the farthest field toward a group of potato pullers. My boots still on, I arrived at the potato patch, knelt down across from a young woman in the sandy soil, and started pulling red potatoes out of the dirt with my bare hands. (I have to tell you, cult-worried reader, that originally I was supposed to harvest asparagus at Tantré Farm, but I had arrived more than a month late.) There were three other young people harvesting potatoes as well. There was only a smattering of idle chatter, and for the most part they were all focused and working hard. The soil was incredibly soft and lush, and the young woman working directly across from me said nothing and didn't even look my way once. Growing increasingly anxious to find out what was going on here at Tantré Farm with Richard and his devoted acolytes, I started to pump the girl across from me with questions. She resisted, but I persisted. It was much harder than pulling potatoes, but I eventually learned that she was an environmental engineering major working on the farm for the summer as an intern, and was there to learn more about organic gardening, soil science, and community-supported agriculture, or CSAs.* She revealed that she and the other interns, as well as the three full-time people, worked nine-and-a-half-hour days, 7 a.m. to 5:30 p.m., Monday through Friday, with a half day on Saturday, 7 to 1. Some, I was told, worked even more hours. They did get paid for their hard work, but she wouldn't tell me how much.† I also learned that Tantré is not a WWOOF-sponsored‡ farm, where workers often just get room and board without a stipend.

We took our harvest back to the warehouse, where it was promptly washed and dried and parceled out for CSA distribution the next day. I could see Richard right where I had left him,

* In a nutshell, people in CSAs pay a monthly fee to a local farm, which in turn supplies them with fresh produce during the harvest season.

† Try as I might time and time again, I never got one person to reveal his or her salary, and only found this tidbit from an article that appeared a year before in the *New York Times*: "Tantré's interns . . . work and often live on the farm for a season in exchange for room and board and a monthly stipend that starts at $500. Such arrangements have become so popular that some states, including Oregon, have reinforced a federal law from 1938 to ensure that farm apprentices are paid fair wages." Christine Muhlke, "Field Report: Will Work for Food," *New York Times Magazine*, September 8, 2010.

‡ WWOOF refers to World Wide Opportunities on Organic Farms, an organization that matches volunteers with organic farms. WWOOF hosts do not pay volunteers for their help. WWOOF organizations usually charge a small fee to hosts and volunteers. This fee helps maintain and develop the WWOOF network. See www.wwoof.org for more details. As I write this, my oldest daughter is WWOOF'ing for two months on a sheep farm in Ireland.

only now he had his tractor with him. I decided to skip more work and seek out his company.

We must have stood there and talked for an hour. I should rephrase that: Richard talked; I just nodded and jotted down notes. Richard Andres was not born into farming, not by a long shot. The details he recounted about his life were, at best, very blurry—he's a soft talker*—and I wanted to learn more, but I am just not the obnoxious, nosy type, so I took what I could get, scribbling furiously. The gist is that during his youth (he is forty-eight, but I understood him to mean some time around his college years) he became a Buddhist monk—*ascetic, monastic, vegan, recovery, healing, overeducated* were some of the noteworthy words that jumped out at me as he described that period of his life. He eventually became a housebuilder, leaving his monk's frock somewhere behind him, though he never did say when or if in fact he did. It was while doing construction on a house near Tantré Farm that he became interested in farming. Later he met his wife, Deb, a schoolteacher, who had been raised on a Minnesota farm, and together they bought the forty acres in Chelsea, Michigan, and started an organic farm, complete with dairy cows, chickens, pigs, herbs, flowers, fruits, berries, and vegetables. Today they have 350 CSA patrons and sell at area farmers' markets, including the big one every Saturday in downtown Ann Arbor.

Then Richard began to speak beyond the facts, revealing his more philosophical side. It was while I was frantically writing down his epigrams that I began to understand why there was such an aura of the charismatic leader about the place—

* If you're a *Seinfeld* fan, you likely know the term "low talker" for this.

Buddhist monk/farmer/philosopher? *Ah So.* He had some serious axioms, too, as they related to organic farming. For example:

"Fertility science is weed science."

"We've got to take commodities out of the realm of food and put them into the realm of the exercise club. It will discipline us all."

"Farming is basically energy maintenance and production; energy stewardship and transference; energy restoration and re-distribution."

"If it's not a closed circle, it's not a farm. These big farms are only half circles."

He then began to wax negative about the neighboring farms and how many of them only grow wheat, corn, and hay, which, he opined, made them "highly dependent on big equipment, which tends to isolate people, who in turn tend to be racist, isolationist, and misanthropic." Then he began to talk about a new project in Ann Arbor he had just become involved with—an inner-city farm, designed after Will Allen's "Growing Power"*—on a seven-acre parcel near downtown and close to the restaurants, with a year-round employment component of people jarring and canning fermented root vegetables. He paused on that for a long time, as if savoring it. Richard is, indeed, a rara avis, and perhaps representative of a new breed of farmer that is emerging across this land, having found their way onto the land and soil via their hearts, minds, and bodies.

---

* Growing Power is a national nonprofit organization and land trust that supports people from diverse backgrounds by helping to provide them with access to healthy, high-quality, safe, and affordable food. Growing Power also provides hands-on training that teaches people how to grow, process, market, and distribute food in a sustainable manner.

Then, with his eyes widening a bit, he put his straw hat back on his head, smiled enigmatically, and said: "You know what? I've got to get this garlic planted, and if I don't do it right now, I won't do it."

And like that, our conversation was over.*

I drifted back up toward the house. It was close to six o'clock in the evening, and most of the farmworkers had quit for the day. There was a sink on the side of the house where the workers washed up or grabbed a drink of water. It was there that I bumped into Richard's wife, Deb. She was with another woman, a CSA patron, I soon learned, who had her daughter along with her as well as her dog. Both women looked to be about forty. Deb had reddish hair, shiny blue eyes, and a flinty voice. She was loquacious, effusive, and amiable to the extreme. I introduced myself, and she took me by the arm and led me around the house.

We walked over to an area behind the gingerbread shed, on the slope toward the bean field, and stood back and watched as the CSA mom put her daughter on one of the two long-roped tree swings and pushed her. It was a major swing, one that would have taken my breath away as a kid. The branch from which the rope hung must have been thirty feet from the ground. After the second push, the CSA mom turned to me and said: "I come here with my daughter once a week. She absolutely loves it here. Three hours is the minimum time we can spend before I can get her into the car and head home." What a perfect advertisement she was; but I didn't disbelieve her.

* It turned out in the end that it would be the only conversation of real substance we would have, he being a hard man to pin down.

We watched the kid swing a little, and then Deb took me across the street to another cute-as-a-button farmhouse. It was perhaps a hundred yards from the main house. There was a barn behind it with a corral full of cows, and two other small dwellings, like one-bedroom shacks. There was a garage across the driveway where five tractors were parked, lined up neatly in a row and glistening like they'd just been delivered from the factory. Deb explained that they had purchased the property from an elderly neighbor fairly recently, and all of the buildings, including the barn where the dairy cows were housed, were in very poor shape when they bought it. Richard, being a master carpenter, fixed them all up in no time. Half of the eighteen workers lived in these various buildings.

She showed me into the house. There was a nifty little kitchen as you walked in, with a Formica-topped table piled high with Ball jars, milk jugs, a clipboard with names, and a jar full of money. "We sell raw milk," she explained, and motioned toward the extra-large refrigerator. She opened it up to show me the half-gallon jars of raw milk stacked tightly on the shelves.

"What's so great about raw milk?" I asked. Stupid me!

"Pasteurization destroys all of the nutritional value in milk. It kills all the beneficial bacteria."

"Beneficial bacteria?"

"Yes. Raw milk is alive. There's a great deal of research that connects pasteurized milk to allergies and even cancer. Raw milk with all of its good bacteria helps digestion. It's got more calcium. Most of our raw milk consumers are weaning their kids off breast milk and onto raw milk."

"This is news to me."*

Next, she walked me toward the back of the kitchen, where an alcove revealed a small laundry room, across from which was a bathroom. Someone was in there showering. She pointed to the stairs leading to the second floor with the comment "There's another bathroom up there if this one is occupied." Then she took me into the living room area and motioned toward a very long couch under two windows against the western wall. There was an acoustic guitar lying on it and a floor-to-ceiling bookshelf on each end.

"If you don't mind sleeping on a couch, that's yours."

"Looks deluxe."

At that precise moment, one of the young female farmhands came strolling unabashedly through the area from the bathroom, before disappearing behind a bedroom door directly across from the couch, not but ten feet as Cupid's arrow flies. She was all covered in towels. I'm sure my face must have given off all sorts of flashing red signals and alarm bells, because Deb instantly said, "There are guys living upstairs. We also have a tent that we can set up outside if you'd prefer more privacy. Sorry, but we're totally maxed out on sleeping space."

"No, no, this will be fine," I lied. I had a vivid image of myself snoring, all sorts of loud farting noises emanating from my backside as a chorus line of half-naked, towel-clad twentysomething beauties high-kicked over my head and through the room, à la the Rockettes . . .

---

* For an excellent, not-too-technical explanation of the sensual and scientific virtues, benefits, and reasons to drink raw milk, see Kristin Kimball's *The Dirty Life: On Farming, Food, and Love* (New York: Scribner, 2010), p. 93.

"Are you sure?" She broke me out of my reverie.

"Yeah, yeah, I'll be fined—I mean *fine* right here," I Freudian-slipped.

"Well, I've got to go help get the dinner on the table. Whatever you need, just ask." And out she went.

I placed my pathetic little duffle bag in an out-of-the-way corner and my rolled-up sleeping bag next to the guitar on the couch, and with nothing to do but think about the dancing legs and the bedroom and the towels ten feet away, I let my eyes wander around the room. Instantly they lit on the copious selection of books jammed cheek to jowl on the wall-to-ceiling bookshelves. What a library! History. Science. Sociology. Psychology. Religion. Philosophy. Bertrand Russell, Shakespeare, Dostoyevsky, Doris Lessing, Guy de Maupassant. And organized, too. I pulled out one book that was bulging toward me as if to say, "Pick me!" and within seconds I was seated and reading *On Food and Cooking: The Science and Lore of the Kitchen*, by Harold McGee.

MAMMALS AND MILK

The Evolution of Milk

> How and why did such a thing as milk ever come to be? It came along with warmbloodedness, hair, and skin glands, all of which distinguish mammals from reptiles. . . .

Suddenly, from out of the same bedroom came another young woman, in a tank top and very short pants. She smiled my way as she passed into the kitchen and then out the door. I watched her traipse across the front yard toward the house, but suddenly became self-conscious and forced my eyes back down toward the page, where I came face-to-face with this highlighted quote:

> Has thou not poured me out as milk, and curdled me like cheese?
>
> Job to God (Job 10:10)

A moment later a great clanging bell forced my thoughts up and out of the book. It was coming from the homestead, and beckoned all to dinner. Suddenly, like rats fleeing a sinking ship, bodies came from all over, stairs and closed doors, as they ran out of the house to the call of vittles. I ran, too.

Inside the main house it was absolute mayhem. On a large round table in the adorable little dining area adjacent to the equally small kitchen and far smaller living room sat large square rectangular sheets of vegetarian gruel, geometrically contrasted by large round bowls of salad and fruit and great bulbous jugs of water as well as exotic-colored teas. And all around it a frenzied hoard of dirt-and-sweat-soaked youth stabbed and grabbed and chewed and gulped. I couldn't get anywhere near the table and so opted to stand on the sidelines, near the front door, and watch. I counted seventeen people, about a fifty-fifty mix of male and female, with two older men in the mix. One of those men, Serafino, had come up from Mexico many years ago. He was about my age and was easily the most aggressive forager at the table, pushing in to get his grub wherever he saw something he wanted. The other older guy, nicknamed Cheez-o, looked to be about the same age as Serafino and me. He had a dark complexion, a shock of thick black hair, and intense coal-black eyes. Cheez-o had been one of Richard's fellow Buddhist monks back in the day. Both of these men, I would learn, worked full-time on the farm, year round. I also learned that another older year-round farmhand had up and quit the day before I arrived, something he did every year about this time in the season. There was another year-rounder in the mix, a thirtyish-looking dude, rakishly thin and lissome, with a long ponytail. He would be the one and only crew member I never would get around to meeting.

What a madhouse! It was easily 90-plus degrees in there, with no movement of air save for what was coming out of people's mouths, and yet the entire crew, with the exception of Serafino and Cheez-o, opted to cram together in the tiny living room and eat on top of one another. Not me. Once the feeding frenzy had abated sufficiently, I sidled over to the table, absconded with a few morsels of home-cooked fare, and skedaddled to the back of the house, where three large picnic tables sat under the cool shade of numerous persimmon trees. Richard was there doing his Buddhist-monk, nodding-off thing.* Cheez-o was also there, talking with Deb about the Pennsylvania Turnpike.

"I hate that road," I interjected. Bad move. Cheez-o shot me a look like I had just spit on his food.

"What's wrong with the Pennsylvania Turnpike?" he growled at me like I had insulted his family.

---

* On each and every occasion that we sat at the picnic tables under the trees, for both lunch and dinner, Richard would spontaneously nod off to sleep, his eyes and body twitching from REM. I am not a snoop, and so I opted to investigate this "disorder" of his on my own rather than ask him outright, "Why do you sleep sitting up?" Here's what I found, a likely reason why Richard liked to sleep in his chair: "Also if you are more upright when you sleep, when you wake up you haven't slept so deeply, and it is easy to wake up quickly and get going." From: "Why do Buddhist monks sleep upright?," BBC, June 22, 2009.

"It's too narrow and crowded. It's dangerous. I was on it last year, driving from Pittsburgh to Gettysburg." If looks could kill, I'd be dead. Perhaps his grandfather had built the road, I don't know, but I called an audible:

"So what's up with the eating scene back in the house? Why aren't they out here where there's something like a breeze to enjoy?"

"I know!" Deb shrieked, as if glad someone had finally made mention of it. "I don't know why this group eats like that. None of the previous crews we have had ate like that. It's an odd bunch this year."

"So most of your workers only work one season?" I interpreted.

"Yeah, that's generally how it works."

"And who does the cooking?" I asked.

"They all do."

"Do they get time off from working out in the field to cook the food?" I was really cranking out the questions.

"Yes, they do. But not the dinner cooks. They work until five thirty and then put the meal together for seven o'clock."

"Do you cook, too?" I turned and asked Cheez-o, who was sitting next to me on my right. I could tell right away that he did not like me at all. Perhaps it was because I just barged into the conversation without introducing myself. Anyway, he was in the act of putting a forkful of salad into his mouth, and when I turned and asked him the question he froze and then turned slowly toward me, his dark eyes flaming balefully:

"Oh yeah" was his response, laced with acidic sarcasm, before he jammed the forkful of food into his maw with a vengeance.

"Everyone loves it when Cheez-o cooks. He's really quite the chef," Deb came in, smoothing it all over as Cheez-o smoldered there next to me.

"And what about the other older guy?" I looked deliberately over at Cheez-o, who by now had simply put me on "ignore." So I turned back to Deb expectantly. She knew whom I meant.

"You mean Serafino?" She nodded at me knowingly. "You must know something about Latin culture. No, he's the only one who doesn't participate in the shared cooking, but we're working on him. One day he'll make a meal."

I could feel Cheez-o's intense eyes burning a hole in the side of my head because he had correctly intuited that I was getting ready to launch an all-out question attack on his person, and as I turned toward him, he preempted me:

"What's your story?" He jabbed antagonistically.

So I told him my story, which elicited a sour frown. Then he hit me with a real gut shot that knocked the wind out of me: "Why are you only interested in harvesting one crop at a time?"

"Well, I . . . I . . . the book . . . that is to say, I was supposed to come earlier and harvest asparagus, but—"*

"Asparagus!?" he scoffed. "That was all done a month ago. If you're trying to harvest one crop at a time, why aren't you out in California on some monocultured mega-farm?"

"No, but I . . . I'd prefer a variety," I sputtered incoherently,

* Cheez-o's question made me reconceptualize the whole idea. I thought that I could simply travel around from farm to farm and harvest one crop at a time at each place. For some farms, like the Stephenses', that only grow one crop, that would not be a problem. But on a farm like Tantré, which is all about diversity, well, I'd have to be adaptable.

totally contradicting my original statement. If we had been playing chess, he would have checkmated me in three moves.

Then he began to pepper me with questions, each one more difficult to answer than the next. But I understood his motive—he was not going to allow me to ask him any questions about himself, no way. Eventually he finished his meal, stood, and shuffled off without comment. He was a dark and mysterious character all right, a regular Pavel Fyodorovich from *The Brothers Karamazov*, and one whom, unfortunately, the scope of this book cannot adequately develop. However, my resolute reader, I will tell you this: I learned later on that he was much older than he looked (not close to fifty but more like sixty), and he lived alone in one of the smaller shacks out behind my sorority house. I was also told that whenever he got to feeling sick he would fast for days until he was better, and on weekends he would keep to himself, holed up in his little home in dissipation.

Deb, forever merry and bright, filled me in on more of the farm's background. She explained how it was really her dad, an old German farmer from northern Minnesota, who had been instrumental in getting them started in farming. He gave them the confidence and guidance they needed. He is still a valuable resource, she explained. Her brother, too. He lives nearby and services all of their tractors, one of the few things Richard is not trained to do. (I saw the brother come and go on a few occasions, sticking strictly to the shed where the tractors were stored. He never came into the house or joined us for a meal.) I revealed to her that I had seen the tractors and was amazed at how neat and new they all looked, compared to the ones at the farm that I had just come from. We both, at that point, looked over at Richard, who was nodding and twitching. She held her hand up to her mouth and whispered behind it to me: "We've got *way* too many tractors. He just likes buying them at auctions."

After dinner, I strolled around the entirety of the farm. The dirt road I'd driven in on made a sharp, 90-degree turn past the main farmhouse, and I strolled down it to the back end of the forty acres, under the bowers of huge hardwood trees—maples, oaks, and walnuts. There were greenhouses, or hoop houses, as they called them, at the far end of the farm, past the bean field and next to a small pond. Both were packed full of tall, ripe tomato and pepper plants. On the other side of the road, at the extreme far end of the property, behind a low stone wall and hedgerows, was a five-acre parcel full of corn. It looked pretty dry and beaten up and weedy, too. While I stood on the edge of the cornfield looking at the crop, I heard an exotic warbling off in the distance. Sandhill cranes! I recognized their call from my days living in Wisconsin. It was a sound that I would hear intermittently for the next four days. I searched up and down for the cranes until dark but could not find them.

That night I lay on top of my sleeping bag, in my bathing suit, drenched in sweat, tossing and turning and not sleeping a wink. It wasn't so much the heat but the humility of the situation, me lying there for all to see, on the couch, shirtless, in my bathing suit, a grown man with a family, coeds lurking behind every door. Nonetheless, I did finally manage to catch some z's until about 5 a.m., at which point all hell broke loose. There was such a racket of clanging pots coming from the kitchen that I sprang up from the couch and ran out to see what was the matter.

Standing in the doorway of the kitchen, I watched Serafino, his back turned toward me, water running full blast, washing the milking equipment in the sink, his hands a montage of blurs. It was all those stainless-steel parts banging against one another—the teat cups, the ten-gallon bucket, and the basin itself. They crashed and clanged and sent shivers down my spine to the tips of my toes. He did not know I was there as I stood for a few moments watching him maniacally washing and rinsing, washing and rinsing the items. I crept back to the couch and tried to go back to sleep. I put the pillow over my head, but it was no good.

At breakfast in the main house at about 6:30 a.m., I found myself standing in the small kitchen, more dead than alive, gawking at a half-dozen youths making their own breakfasts. It was a do-it-yourself operation and a lot less hectic than the scene at dinner the night before. There were boiled eggs, cut fruit in bowls, cereal, English muffins, sliced bread, butter, jars of jam, and a quart of yogurt on the kitchen table. Someone had made a large pot of oatmeal that was sitting on the stove with a big wooden spoon sticking out, so I helped myself, adding a little yogurt and fresh fruit to a small bowl I found in the cupboards. I ate standing up, uttering an occasional "good morning" whenever a new face appeared.

Outside and in, it was another insufferably hot day.* Somewhere out there, the sandhill cranes were chortling back and forth to one another as if taunting me to find them. I could hear

---

* I learned later that the average temperature during July 2011 nearby in the city of Detroit was hotter than any single calendar month on record since data had been collected back in 1874.

them in between the subdued conversations around me. The house was filling up, so I slipped back out the front door and ambled across the street to see if I could spot the cranes in the wheat fields. No luck. It was almost seven o'clock. Even with the sun low in the sky, I was beginning to get that wilted feeling. I tried to put the thought of working a full nine-and-a-half-hour day out of my mind, but the mounting drops of sweat on my brow just wouldn't let me do it. When I turned around, Richard was walking toward me with a Buddha smile.

"Ready to work?"

"You betcha!" I didn't sound very convincing, notwithstanding the Spanky jargon.

"We're picking squash and lettuces today for tomorrow's CSA. How about that?"

"I thought we were picking green beans," I replied.

"That's tomorrow."

Cheez-o's comment came to the fore, and I understood that going with the flow was once again *condicio sine qua non.*

"Then squash and lettuce it is."

He paired me up with the same girl I had worked with the day before on pulling potatoes. She was an enigma, that one, wrapped inside a brown paper bag. Petite, blond, fair-skinned, blue-eyed, flat affect, with a weird glimmer of . . . something, but I'm not sure what. Try as I might to get her to talk about her Tantré Farm experiences, she just wouldn't respond, but instead squeaked in low tones, partially audible, about her goal to one day design green roofs for city buildings. Green rooftops gave my quixotic mind something to think about beyond my present morass, and as I harvested I indulged in visions of futuristic cities, like scenes from the movie *Sleeper*, full of colorful monorails and stylish hovercrafts, exotic green spaces and fantastic solar collectors, miles of meandering movable sidewalks and radical, loop-the-loop zip lines, none of which would ever come to be.

But the inspirational eventually gave way to the irritational as the nettling vines of the squash plants began to take their toll. I had gloves on, which saved my hands, but my short pants were no protection from those discomfiting little needles that covered all parts of the plant except the yellow fruit itself. I was equally flummoxed by just how evasive those yellow schnozzolas could be. Plants from whose vines I had harvested in one row would reveal unpicked fruit that I had overlooked from the point of view of the other row. I racked my brain to try to understand what the evolutionary advantage was for squash to nettle and hide from their patrons at every opportunity even though they benefit greatly from the harvesting of their fruit. Squash is one of the famed "three sisters" crops that Native Americans used to grow together. The first sister crop, corn, with its tall and sturdy stalk, would provide support for the second sister, the climbing bean, which in turn would provide additional shade for the third sister, squash. Squash vines provided cover to the other two to limit weeds, while the beans provided the all-important nitrogen-fixing bacteria that would help all three grow and multiply.*

---

* I asked a botanist this question, and his answer was simply that the nettles helped discourage insects and weeds from eating and competing with the maturing plant. As for the hide-and-seek quality of the fruit, he explained that squash leaves are thick and add shade; otherwise the fruit would grow too quickly or dry out.

Within a couple of hours, we had picked four rows of squash, no small feat. It was only a little past nine in the morning and I was already starting to sag. I staggered back toward the warehouse to drown myself with water, my legs on fire from the nettles. My coworker, the faraway blonde, whom I decided comported a phlegmatic disposition that put her in the same taxonomic phyla as the Mollusca and Porifera, had gone straight away to the lettuce patch without even a sip of water or a revivifying ten-minute break. Not me. After a good break and a thorough drenching at the spigot—head, legs, and stomach—I stood there watching her. She was all crouched down in the lettuces, methodically cutting little bouquets of arugula and bundling them with rubber bands. Submitting to my fate, I took a deep breath, rocked once, twice, thrice, then went and joined her once again.

She showed me how big the bundles should be, and how to hold them and tie them with the rubber bands. I actually enjoyed the job for two reasons: 1) the odor of arugula is a heady, invigorating smell, evocative of wine and fine dining; 2) the tying of the bundles is exactly the same as tying hair in a ponytail. The knife was a little dull for the work, and try as I might to lop off the leaves at their bases, I always managed to mangle the leaf heads terribly. In comparing bundles lying in the crate, hers presented far more deliciously and floridly than mine, as her nimble little fingers had deftly arranged the redolent leaves into fluted, come-hither bouquets; mine looked like masticated vegetable matter. Nonetheless, she made no comment, as together we carried our harvest back to the sorting area.

Halfway to the warehouse I cavalierly asked her: "So you ever eat arugula pesto?"

She just shook her head, her little retroussé nose pointed at the sun. I didn't bother to inquire further.

It was after eleven o'clock, and the sun and heat were laughing hysterically at me. I was pretty much a zombie by this point, and, in fact, began to grow more into that role throughout the rest of the day. Back inside the packinghouse, the crew was on fire, filling boxes and washing produce, rushing around and calling back and forth to one another, their eyes wide, their hands and feet in constant motion. One very pretty intern with an extraordinarily tall top-knot was practically dancing from cooler to bench to sink and back. I stood there slack-jawed, dumb, numb, functionally idle. I didn't even try to insinuate myself into the work zone. Then, from behind me, the rakish, ponytailed manager dude appeared on the scene, riding up on a cart, all enthusiasm and unction, and, in my quasi-comatose estimation, suspiciously dry and sweatless.

"Hey, guys, we've really got to pull those weeds over in the corn patch. Who's game?"

I don't remember exactly how the "volunteering" went down, but I do recall a sort of reverse rendition to the Little Red Hen, at least the first part of it anyway, where I was the naysaying Little Red Hen and the enthusiastic crew was the all-affirmative responding dog, cat, and pig.* The next thing I knew I was out in

* Once there was a little red hen who decided to grow some wheat. She asked the other animals in the barnyard, "Who will help me plant my wheat?" "Not I!" said the dog. "Not I!" said the cat. "Not I!" said the pig. "Well, then," said the little red hen, "I will do it all by myself." And she did.

The wheat grew tall and was ready to be harvested. The little red hen said, "Who will help me pick my wheat?" "Not I!" said the dog. "Not I!" said the cat. "Not I!" said the pig. "Well, then," said the little red hen, "I will do it all by myself." And she did.

the middle of a life-threateningly hot cornfield, hoeing. This now . . . this was beyond stupid. This was pathological. At one point, feeling as if I was going to faint, I knelt down in the sandy soil to gather myself. Suddenly a pathetic whimper and then a zoological roar erupted from my lungs as my knees began to burn. The sandy soil was so hot it was scalding my skin at the knees! I jumped up, twitching and convulsing. And who was right there in front of me, not but ten feet away, not even bothering to glance at me? You got it—Mollusca Porifera. She was dipsy-doodling along, not a bead of sweat on her brow or a hair out of place, a little gloating smirk tipping her lips up at the ends, happily hoeing along. *Arrrgh!* I gritted my teeth and continued down the row, hoeing like a madman until I hit the end fifty yards farther on, at which point I collapsed under the shade of the trees and lay there for the next twenty minutes stewing in my own emasculated juices, watching that MF MP luxuriating out in the diabolical heat.

Next, the wheat had to be taken to the mill and ground into flour. "Who will help me grind my wheat?" asked the little red hen. "Not I!" said the dog. "Not I!" said the cat. "Not I!" said the pig. "Well, then," said the little red hen, "I will do it all by myself." And she did.

Finally, the flour was ready to be made into bread. "Who will help me make my bread?" asked the little red hen. "Not I!" said the dog. "Not I!" said the cat. "Not I!" said the pig. "Well, then," said the little red hen, "I will do it all by myself." And she did.

The bread smelled so good when it was baking. All the animals' mouths started watering as they gathered around hoping to get a piece of bread. "Who will help me eat my bread?" asked the little red hen. "I will!" said the dog. "I will!" said the cat. "I will!" said the pig. "No!" said the little red hen. "You did not help me plant my wheat. You did not help me pick my wheat. You did not help me take it to the mill to be ground into flour. And you did not help me make my bread. I did it *all by myself*! And my little chicks and I will eat it all by ourselves!" And they did!

It was while sitting there in the shade, as insight slowly began to replace heat prostration, that it became clear just how vital youth was to the whole enterprise of agriculture. I went back in my mind to my football days of two-a-day practices in the extreme heat, and to that summer I worked as a drywaller, double-rocking closets on top floors in the noonday humidity . . . In sum, she was no maniac. I used to be just like her. I worked and played without complaint in the heat, indulging in my capacity to withstand physical pain. That's what youth does. Like all sports and activities in life that test the limits of human endurance, they are best and perhaps only performed by youth. When the lunch bell finally clanged at one o'clock, rigor mortis had nearly set in.

Under the persimmon trees, it was just Deb and I for lunch. We sat there and talked for the longest time about the state of education—long after the crew had finished its lunch and resumed its work. I did not intend to string things along; it was almost as if she could tell I had had a miserable morning and that I needed to chill for a while. Leaning on my elbows at the picnic table in the cool shade, I listened to Deb give extraordinary examples of her elementary classroom experience, lessons that worked magically and those that failed miserably. Together we began to offer our vision of alternative education programs, especially for unmotivated and uninspired students, that would incorporate farming directly into the curricula. Let young people learn how to learn by working the land, growing and raising food, tending fruit trees, composting, managing livestock, selling and inventorying produce, learning through real experience. "Farms into Schools" was the title I later came up with.

After our idealistic conversation eventually resolved itself of its own volition, Deb headed off to tend the chickens while I wandered back down toward the warehouse to see if I could be of use. As I came over the hill, past the chickens, flowers, and berry bushes, the beauty of the rows of herbs and vegetables revealed itself to me as an eloquent, five-acre script. Growing there in long, ornate lines, like a living text, the varied rows of herbs and vegetables, each plant a letter, each row a phrase, presented a transcendent form of communication. All the glorious shapes and flavors, colors and odors, tastes and varieties expressed in line after line the highest form of human achievement and human passion and human desire.

I meandered slowly down to the front of the warehouse and leaned against the side of it, where I stood and watched the workers crouched down in the rows, peacefully picking produce. I stood and watched for the longest time before I once again became aware of something extraordinary: all of the workers who passed by me did not seem to see me. It was apparent that workers on a farm, especially those hard at it during the harvest season, do not react to idleness, in the same way that predators do not react to a dead or motionless body in the animal kingdom. Movement, action, work is the only recognizable profile; anyone who is unemployed or inactive out in the field ceases to exist.

I pushed away from the wall and began to move around again. No sooner had I begun to move than Cheez-o, who was working with one of the younger workers replacing drip lines in the newly planted rows of lettuces, motioned at me.

"Hey you! Grab that rod over there and straighten these hoses along the beds."

"My name's Rick, not Hugh," I corrected him.

He didn't acknowledge the joke.

There were three hoses that had been laid down in the thick humus, but they were all twisted and zigzagged and bisecting one another. I don't know why Cheez-o and the young farmhand were in such a hurry, but they were hustling along the beds, laying lines as fast as you could say "Bob's your uncle." In fact, Cheez-o, every time I watched him out in the field, was working at breakneck speed. They all worked at breakneck speed. It was unsettling for me, something I never did get used to.* So I picked up the rod with an odd little hook on the end that he had pointed to and began to straighten the lines. It was amazing to me that they had a tool for that very purpose. Nonetheless, it was not as easy as it seemed. For one, I didn't want to mush the newly planted seedlings by placing the hoses too close to them. And two, it was really hard to walk along the edge of the rows and not step on the produce in the adjacent beds with my size 11½ boots. But it was work, good work, relatively light work, and people all around were at least "seeing" me again.

It was somewhere close to quitting time when I heard the cranes start to chortle. They were very close this time, and since I was done straightening the hoses, I dropped my Cap'n Hook rod where I'd found it and hoofed it back toward the sound of the cranes, in the direction of the sorority house. As I crossed the road past the main farmhouse, I heard a very loud chortle,

* Having worked as a teacher for most of my adult life, with some odd jobs in hospitals and nursing homes before that, I was not familiar with this galloping pace of work. Hurrying, in my mind, was synonymous with haste. But harvest season on any CSA farm demands prodigious effort on the part of the workers; otherwise the product rots on the vine.

and two strides later there they were in front of me, the sandhill cranes. There were two of them, down in the back pasture, about a hundred yards away. What exotic animals! They looked like they were escapees from the zoo—four feet tall, bright rusty brown feathers, stilt-like legs, bulbous bodies, and gracefully long necks. They moved deliberately through the tall grass. One suddenly gave a great chortle and then disappeared into the hedgerows. These birds held some meaning, and I made a mental note to find out what that meaning was.*

That night I drove into Ann Arbor to have dinner with an old friend and a former colleague whom I had recently found on Facebook. He was living in Ann Arbor and teaching at a middle school. He and I had started our teaching careers together some twenty years before in a public high school in rural New Hampshire. I hadn't seen or spoken to him since then. That had not been a very happy time in my life, and I didn't like to think back on it if I could help it. The job was an unqualified disaster, and I ended up quitting before the school year was even out, vowing never to teach high school again. Meanwhile, our first child had been born. We had no money and no health insurance and we were living in the middle of nowhere. But things worked out in the end, so there are no real traumas associated with it.† Anyway,

* Cranes represent many things in many different cultures, like honor, loyalty, and peace; longevity, nobility, and spirituality; health, happiness, and recovery. The latter word is, in my opinion, why they are there, at Tantré Farm.

† Eventually I took a job with the Department of Defense Education Association, and we moved overseas to Belgium and later to South Korea,

I met him and his girlfriend at his house on the outskirts of the city, and together we drove downtown for some pizza and beer.

The next morning, my tongue all swollen, I was out in the bean field with Serafino, sitting on top of a five-gallon bucket, a straw hat on my head and a bottle of water by my side. I picked green beans all the day long. Buckets and buckets and buckets of green beans.

As Michael Pollan wrote in *The Omnivore's Dilemma*, "All life depends on nitrogen; it is the building block from which nature assembles amino acids, proteins, and nucleic acids; the genetic information that orders and perpetuates life is written in nitrogen ink." Beans . . . legumes . . . *frijoles, judias verdes* . . . green manure . . . is the quintessential cover crop and the only plant in the kingdom that replaces the nitrogen taken away by the growth and harvest of the other plants. The beans themselves don't fix the nitrogen; rather "nitrogen is actually fixed by a bacterium living on the roots of the legume, which trades a tiny drip of sugar for the nitrogen the plant needs," as Pollan explains it.* And because of their dire importance to the soil health of the nonindustrial farm, I thought I should pay homage to this precious crop by picking it.

And the best and only way to do that is to sit on top of a five-gallon bucket and pick and pick and pick. And that I should pay homage to the magical bean by picking alongside the only Mexican working at the farm seemed truly divine—beans and

---

where I ended up teaching, yes, high school to army brats. I actually enjoyed it.

* Michael Pollan, *The Omnivore's Dilemma* (New York: Penguin Books, 2006), p. 42.

Mexicans having an undeniable affinity. Serafino and I hit it off almost instantly, not only because of our similar age, but also because we shared so many other similarities, like our Latin blood, our Roman Catholic upbringing, our family orientation, and our joyous immaturity, to name but a few. For two days and nearly eighteen hours, we were like peas in a pod as we picked up and down and sideways the rows of beans, occasionally stopping to empty our buckets. He spoke almost no English, and my Spanish isn't very good, but we managed to communicate at an extraordinarily profound level.

Por ejemplo:

Ricardo: Qué piensas de la Pamela Anderson?

Serafino: Pamela Anderson no me gusta. Tiene demasiadas bubis.

Ricardo: Ah sí, mejor la Eva Mendes?

Serafino: ¡Mucho mejor! And jou like watch Los Tres Estooges?

Ricardo: Soitainly. *Y cuál prefieres, los años con Shemp o con Curly?*

Serafino: Con Shemp mucho más.

Ricardo: Yo también. Pero, mi programa favorito de todos los tiempos fue Hogan's Heroes.

Serafino: Bueno, muy bueno! Sergeant Schultz es muy simpático.

Ricardo: Me encanta Sergeant Schultz; pero el mejor personaje es el General Burkhalter.

Serafino: Sí sí, que es un magnífico bastardo!

I avoided stupid questions, like asking Serafino about his H-1B visa status or who his dentist was or where he went to college or how much money he made. I did ask him where he was born—*Oaxaca*. Like me he was married and had only daughters—*trés*. I didn't ask where they lived. He'd been working at Tantré Farm, full-time, for the past ten years. And what a worker! He could pick beans like a house afire. And I could tell he was intentionally going slowly just so that we two could stay relatively near enough to talk. Beans are not an easy crop to pick. Those skinny little demons are incredibly elusive, blending in with the branches or getting lost in the foliage. Bushes that I'd thought I'd picked clean would suddenly come alive with green beans once the bush was pushed forward or pulled back aggressively. But the work is gentle and meditative, albeit tough on the back and on the monotonous side.

For two full days we picked green beans out in the field, under the molten rays of the summer sun, rows and rows of beans. And the more rows I picked alongside Serafino, the madder I grew inside, thinking about those charityless, virtueless, and benevolentless shitheads who have spread about this glorious land a melodyless song, a giftless song that accuses the immigrant of stealing their lunches—when in fact they are picking, packing, and purveying them.* Millions of immigrant workers—

---

* In an in-depth report titled "Replacing the Undocumented Work Force," about the cost and impact of replacing our vital workforce with people from south of the border, David A. Jaeger, writing for the Center for American Progress in March 2006, has this to say in his concluding remarks:

"Many questions in the debate over the role of undocumented workers in the U.S. economy are difficult to resolve, yet some facts are known. The economic analysis presented in this paper illustrates that unemployed native-born Americans would be highly unlikely to fill all of the jobs currently held by undocumented workers, as many opponents of earned

men, women, and children—ignorant, poor, yet so ripe with hope and determination and humility, even while bent over at the waist, picking America's crops, servicing America's insatiable appetite, shouldering the heaviest and most dangerous loads, not so much for themselves, but for America, daily, joyously, like Whitman's song: "A song for occupations! / In the labor of engines and trades and the labor of fields I find / the developments, / And find the eternal meanings."

I found the eternal meanings even more clearly at the end of the second day, when Serafino and I and some of the others yanked out all the bean plants we had picked clean—countless rows, forty yards in length—piled them in heaping mounds onto the back of the truck, and dumped them in the cow pasture as bovine fodder. The great piles of beans stood in the middle of that muddy pasture, twice as tall as the cows themselves, as a testament to the power and force of the farmworker. And as we stood contentedly watching the cows jumping at the pile, I had to ask him:

Ricardo: Cuantos días para las vacas a comer la pila?

Serafino: Dos días.

Ricardo: Dos días!? Holy cow!

Serafino: No, no, farty cow.

---

adjustment programs argue. The skills required by most of the jobs that undocumented workers hold are substantially lower than the skills possessed by out-of-work natives. Moreover, in certain high-immigration states, the contribution of undocumented workers is substantial, and their removal would have a large impact on the state's economy. Demographic trends in age and education of the current American population indicate that rather than reducing our reliance on foreign-born workers, the United States may in fact need to expand their numbers to keep pace with the demands of the economy." I can't help but laugh out loud at that last statement.

After dinner that night, while checking my e-mail on the community computer in the tiny living room, I met the last farm-worker/intern I had not had a chance to meet (the exception being that ever-evasive pony-tailed manager). I noticed him there behind me on the couch, reading David Foster Wallace's *Consider the Lobster.* Totally unprovoked, he started talking to me, revealing that he had recently walked away from his graduate fellowship in neuropsychology at a large research university because he had lost his passion. He had drifted around the country for a while before ending up at Tantré Farm. The monologue picked up steam as he described fits of depression and self-doubt, then simmered back down again as he began to describe his new love of digging in the soil, kneeling on the ground harvesting herbs, and even the smell of the crops, particularly the onions and garlic. All were a sort of spiritual healing for him. I listened patiently, with an encouraging bent to my head. Then the monologue ended, and we sat in awkward silence. So I asked him how he liked the book. He said it was very impressive writing, especially the style and intelligence, but he couldn't reconcile Wallace's suicide. Unable to think of anything to say, I blathered: "Yeah, well, literature isn't about the author's innovations or intellect; it's about the reader's ecstasy and fascination. The problem with David Foster Wallace, besides his pretentious three names, is the fact that he never played in a marching band."

He burst out laughing. "What is that supposed to mean?"

"I have absolutely no idea, but I just know I'm right."

He smiled at me with his mouth wide open and his eyelids blinking like they were communicating some sort of coded message. I stood up and wished him good luck. He stood, too, shook my hand, and thanked me profusely for lending an ear, before revealing yet another morsel about himself—he hadn't really spoken to anyone at the farm since he'd arrived a month and a half before. And as I walked toward the sorority house, his words now images Rolodexing through my mind, I suddenly came to an abrupt halt, one particular image freeze-framed in high definition—that of him kneeling in the soil: *These organic farms are the churches and cathedrals of our time,* I thought to myself just before one of the sandhill cranes let loose with a wild warble.

I left the farm for good early the next morning. I drove into the center of Ann Arbor, ate a hearty breakfast at a diner, and then walked over to the farmers' market located a block from the center of the city. It's a year-round affair, with a permanent T-shaped roof structure. For me it was reminiscent of the open markets in Europe. And even though it was pouring down rain, the place was jammed. Tantré Farm had the premier spot, right at the entranceway. Richard and about a half-dozen crew members were busy selling their gorgeous produce to a standing-room-only crowd. After taking a tour of the market, I managed to spend a few last minutes with Richard before shaking his hand and bidding him and the others goodbye. I felt like there was so much left unsaid and undone; but as we shook hands and looked directly at one another, all those left-unsaids and left-undones didn't amount to a hill of beans juxtaposed with the fact that I had worked and harvested to the best of my ability on his farm.*

* Shortly after leaving Tantré Farm, I found out from Richard that he had purchased a seventy-acre parcel that would more than double his acreage. He intends to expand the operation in the near future.

Blueberries

WILTSE FARM

CONSTANTIA, NEW YORK

JUNE–JULY 2011

It took only about ten minutes at Wiltse's U-pick blueberry farm in Constantia, New York, for my musical wife to burst into song:

I'm a blueberry!
Oompa loompa doompety doo
I've got a perfect puzzle for you
Oompa loompa doompety dee
If you are wise you'll listen to me
What do you get when you pick your own fruit?
Fingertip stains with some blueberry loot
Oompa loompa doompety do.

(She's got a lyric for every occasion, that lady.)

The world-record blueberry, found in Kerhonkson, New York, by Zachary Wightman, weighed a whopping 8.0 grams (about the size of an average crabapple). Presently at Wiltse's the largest blueberry ever picked was 6.2 grams, but Dennis Wiltse, the owner, and his entire team are sure that an 8-plus gram monster is out there somewhere in the berry patch. And that was my mission, to find that monster.

Wiltse's is about thirty miles due east from my home in Oswego, New York. My wife works with the cousin of the owner, and through the years she has never failed to regale her office mates with blueberries during the harvest season, freshly picked from her cousin's farm, so that's why I decided to harvest there. We went twice to U-pick, my wife and I, both times on a beautiful weekend day in the summer, and then I returned by myself twice more to join the incomparable cast and crew to help and harvest "for real." What a delight.

"There's not a blueberry picked that we don't sell," Dennis Wiltse boasted when I first got a chance to talk to him about his farm.

What was so special for me about Wiltse's was the fact that I continued to pick blueberries before, in between, and after my adventures to far-flung farms across the country. It was harvest as recreation, in the true sense of the word. And forget about all that organic, chemical-free, all-natural, biodynamic, polyculture, sustainable-farming folderol and fiddle-dee-dee. Wiltse's is about one thing and one thing only—blueberries. Eight acres of row upon row of large, healthy blueberry bushes, approximately eight thousand all told. Dennis inherited the farm from his father, Harold, who first bought the land (about fifty acres) back in 1965, and planted his first acre of blueberries three years later. The story goes that Harold Wiltse died of a heart attack while out in the blueberry patch, like Vito Corleone, in May 2002. Dennis doesn't like to talk about it, but what he does like to talk about are his nineteen cows—check that, twenty:

"We just had a new calf born. Cute little bugger. You'll see her if you walk back to the end of the patch, hanging close to her mom. She doesn't like to come out of the woods."

All the surrounding faces frowned the minute he mentioned the cows. One crewman scoffed: "Yeah, and they get out at least once a week and eat the blueberries."

Dennis recently clear-cut a wooded, eight-acre section on an adjacent stretch of land, but not for blueberries. It was for his cows. And these aren't your everyday, run-of-the-mill cows, either; they're mixed cattle—for example, longhorns, Holsteins, Charlet, and Black Angus. He raises them for his own table but says he's looking to sell some. Occasionally you see them wandering out of the woods to the north of the clear-cut, where they stay most of the time, to munch on exotic herbs and mushrooms. But it's an odd sight to see them in that clutter field, especially the longhorns, grazing among the rusted water heaters, engine parts, and tractor tires. Who knows how long it will be before he turns it all into real pastureland. No hurry. The cows look quite content.

To the south is a busy road, County Route 49, which runs east to west and connects Utica, New York, to Lake Ontario. Traffic whizzes by at high speed, night and day—people on their way to and from Oneida Lake. Dennis has a little blueberry stand next to the road that reminds me of the lemonade stands that the Little Rascals used to vend from, only *most* of the words on Dennis's sign are spelled correctly.

There's a tumbledown farmhouse with a porch on the side where the meet-and-greet and sorting and weighing takes place. To be honest, the house would not be out of place in Appalachia, with its woefully peeling paint and sagging roofline and cluttered windows. But to be fair, *most* of the house is sided. It sits at the very edge of the blueberry patch like a tide-imperiled sand castle. There is a cluster of tables and chairs in the grass next to the house, and one of those swinging chairs is a veritable grown-up swing set. To the east and north are thick, eastern lowland forests. It is close to a large swamp to the north and is only about a mile from the shore of New York State's largest lake, to the south—Oneida Lake.*

There is a magnificent cast of characters that congregates on or around the porch at the Wiltse Farm, day in and day out, during the summer months. Dennis himself looks like Rasputin. His long, gray-blond hair is parted in the middle, and so is his long blond beard, which reaches down in a stream of uncombed fury (or is it furry?) to his sternum. He's in his late forties. You can't see his mouth for the mustache and beard, but it doesn't matter because his flashing blue eyes do most of the talking. He's tall and likes to wear T-shirts and jeans and lemon-yellow rubber Crocs. The tips of his fingers are indelibly stained indigo blue.

The days that I was there he had four people helping him out: his mother, Edna, Adriano, Fred, and Flo. His mom is a petit bourgeois sort with bobbed hair the same color as Dennis's. Flo was in her late thirties, perhaps. With bent head, white-rimmed glasses, and cigarette in mouth, she worked like an android. Occasionally she would spontaneously launch into stories about her recent trips to the casino, including her latest "winnings," some of which I tended not to believe, but it doesn't really matter, so long as she believed it. Then there was diminutive Adriano, who is Dennis's foil. He is perhaps in his sixties, pint-size with short gray hair and a beard, birdlike legs and neck, glasses, and a baseball cap. When he talks he sounds like he's just inhaled a lungful

* Oneida Lake is the largest lake completely within New York State's boundaries. It is approximately eighty square miles.

of helium. He rolls his own smokes, jacks his shorts way up to his navel, and giggles like a leprechaun. I think he can stand shoulder to shoulder with some of cinema's legendary sidekicks, including Barney Fife, Ed Norton, Mr. Smee, Festus Haggen, Waylon Smithers, Baba Looey, and Boo Boo. Best of all, he's an echolalic: his interjections are like the blueberry farm's own punctuation marker. As for Fred . . . loyal and trustworthy Fred is right between Dennis and Adriano in age and height, with his perfectly smooth and round face and perpetual upbeat and pleasant disposition. He's the meeter and greeter of the farm. For example, when a person is finished picking and ready to check his or her berries on the scale, he is the one who weighs, pays, and okays it all:

"Did you leave any blueberries on those bushes? Looks like you about picked 'em all. Ha ha ha. Okay, okay, let's see what the scale says here . . . Okaaaay, look at that! That's some nice pickin', yes indeed, nice pickin' . . ."

"Some nice pickin'," echoes Adriano nearby.

The first day that I worked on the farm without my wife, I stayed on the porch all morning sorting blueberries. It's a simple process. Dennis takes a five-gallon bucket of freshly picked blueberries and loosely fills dozens of quart containers set up on top of the picnic table situated in the middle of the meet-and-greet porch. The sorters (there were four of us that day: Flo, Fred, Dennis, and me) go through each quart container and remove the bad blueberries, being careful to fill each container as high as possible before loading it onto an eight-quart flat to be sent out to the wholesalers. Repetitive yet hypnotic work, but no muscle cramps or shoulder aches. You can stand up or sit down, change positions, use a different hand, lean your head this way or that way. I did it all. There was lots of palaver, mostly about 9/11 conspiracies, recent debaucheries, infidelities and breakups, and UFO sightings. I just sorted and listened.

"See these?" Dennis held a little white seed, about the size of a BB, in his palm.

"What are those?" I asked. "I keep finding them."

"Mummy berries. It's a fungus. They're extra bad this year."

"Is that your main pest?" I stupidly asked.

"The *main* pest, are you kidding me?" He hopped over to a little table and pulled out a bunch of papers stapled together and began to read: "I got ripe rot, root rot, gray mold, stem blight, not to mention blueberry maggots, blueberry stem borers, Japanese beetles, leafrollers, leafhoppers . . . But I'm staying ahead of them this year. Spraying the shit out of 'em. Little bastards never learn: you don't mess with Dennis's berries."

What was surprising for me was to see how many foreign-born families and elderly folks came to pick the blueberries. I would venture to say that they were the majority of pickers, both for pay and for pleasure, but mostly for pay. There were Ukrainians, Russians, Poles, Laotians, Peruvians, Puerto Ricans, octogenarians, and others of indeterminate nationality. Dennis pays fifty cents a pound for pickers and charges $1.25 per quart for U-pickers.

You could tell the pay pickers from the U-pickers because the former all brought their own buckets and belts and bottled water. They'd make a little small talk with Dennis and the crew, but within minutes they'd get right down to work. It gladdened my heart to see all these happy, healthy people, most of them speaking another language, working together, laughing together, out in the fresh air, picking blueberries. What zest!

Several hours later they'd return to the porch and cash in *half* of their take and take home the rest. But I didn't understand one thing: "What are they doing with all those blueberries that they keep?" I asked Dennis after observing this several times. "That's a lot of blueberries to take home."

"They probably make pies and muffins and freeze the rest to eat during the winter."

"Maybe they sell 'em," I offered.

"Yeah, maybe they sell 'em?" Adriano said in support.

"Some people do, but where are they going to sell 'em?" Dennis disagreed, shaking his head.

"Friends, relatives, at the farmers' market, or on the side of the road out in front of their house."

He didn't respond, just kept on sorting blueberries, but it made sense to me. (I later found out from my wife that Dennis's own relatives had been picking at the farm and selling their harvest to friends and relatives for years.) It made perfect sense: If they're paying Dennis only $1.25 a quart, and blueberries sell for $4.50 a quart at the supermarket, what's to stop them from selling them on their own somewhere down the line?

"Blueberries are like the number one food on the healthy food list," I blurted out, apropos of everything. Again Dennis hopped over to his little table and pulled out another crinkled sheet of paper and started reading again:

"It's got antioxidants, which help cells grow. Fights aging and aging-related diseases like Alzheimer's. S'got anti-inflammatory properties. Inhibits cancer. Helps the urinary tract. Protects against stroke and heart disease. Improves and repairs vision. Counteracts high blood pressure, diabetes, and metabolic syndromes. It's like a miracle drug," he finished, shaking the papers and slapping them.

"It's a miracle drug?" Adriano reiterated.

The first chance I got to pick blueberries finally came late in the day when there was a bit of a lull. I finally looked over at Dennis and said, "You know, I've gotta get out there in the blueberry patch to test my picking chops."

"Sure, sure. Get a bucket and go."

So there I was, out in the blueberry patch, a two-gallon bucket dangling from a belt they gave me lashed around my midsection. The reason for the bucket and the belt is that you then get the use of both hands. The bushes were human height, so there was not a lot of bending or stooping, just reaching. Dennis instructed me to go over into what he called "the Jungle," an area of bushes on the eastern side of the patch close to the woods that was completely overgrown; it was for that reason that there were lots of berries still left on the bushes. He was convinced an 8-plus gram monster lurked somewhere in there.

The happiness expert Mihaly Csikszentmihalyi has proven that the difference between enjoyment and pleasure is the fact that enjoyment involves achievement and pleasure does not.* He

* For a comprehensive look at happiness research, check out Mihaly Csikszentmihalyi's (pronounced "Me High Chicks Send Me High") *Flow: The Psychology of Optimal Experience* (New York: Harper Perennial, 2008),

explains that when happiness flows, "[t]he ego falls away. Time flies. Every action, movement, and thought follows inevitably from the previous one, like playing jazz. Your whole being is involved, and you're using your skills to the utmost." Picking blueberries is flow happiness, especially of a sun-drenched, blue-skied, summer-sweet day. By yourself, using your body as a harvesting mechanism, you feel that all is right with the world, especially since the faster you pick, the more money you make. And what better product to be picking? On the downside, the bucket just doesn't seem to want to fill up very fast.

So within minutes, I found my fingers working the bushes like a prestidigitator. One technique I picked up on right away was to avoid going after single berries and to be on the lookout for clusters. Clusters are key. The problem is, within every cluster there are a bunch of losers, such as greenies, overripes, mummy berries, bee-eatens, etc. You lift your cluster delicately off the branch, then you delicately cull the bad from the good and gently drop them in the bucket. There is also a pride factor at work when picking. I learned this at the sorting table. When Dennis poured those buckets of berries into the quarts, those buckets were the work of individual pickers. And Dennis knew exactly who had picked which bucket by the quality of the product that came tumbling out. Good pickers had mostly good berries. Bad pickers were full of losers. So if a greenie got in the bucket, I'd take the time to remove it.

It took me about forty-five minutes to fill the bucket. As I was getting ready to return to the porch with my haul, I became

---

or read more at http://www.answers.com/topic/mihaly-csikszentmihalyi #ixzz1WWrvjoKZ.

aware of a family of pickers in the row near mine. They were talking about rent money and bills, and how long they would have to pick in order to make a dent on their debts. I could hear a little boy with them, too, who was trying not to be a burden, but he was just too young and too rambunctious and too delighted to be in such a magical playground like this to be of any help to them, and as I walked out of "the Jungle," I came face-to-face with him. He held his empty bucket up toward me as if to say: "Look, we've got the same bucket."

"I'll trade you," I said.

I held my twelve-pound bucket of berries out to him, and then gently exchanged his empty bucket for mine. He held it with both hands, and then went running to find his father.

As I disappeared back into the Jungle again, I could hear him exclaiming with glee:

"Look what the man gave me!"

When I finally returned to the porch with a full bucket an hour and twenty minutes later, I could tell by the looks on their faces that Dennis and Fred and Flo were not in the least bit impressed. So I told them why I had taken so long.

"The little Puerto Rican kid?" Dennis asked.

"I think so," I added.

"Good," he smiled with a nod.

"Can you weigh me, Fred?" I asked.

Fred took the bucket, and with his usual glad tidings, put it on the scale:

"Well, that's a pretty full bucket for a rookie. Yes, sir, looks like nice ripe product inside there, too. Okay, okay, what've we got here . . . ten and a half pounds. You want to take these home? I'll bag 'em right up," he asked.

"Not all of them. Half the bucket."

"No no, take them all."

"Ten pounds of blueberries is more than we have room for."

I took home five pounds, and by the end of the week they were all gone.

I returned six days later, late in the morning, determined to pick a hundred pounds of berries. I picked only about half that. It took me close to five hours, with a few breaks in between. The payout was $23.25. I was spent. My arms were tired, my back was aching, and I was all scratched up. Because it was late in the season and most of the bushes out in the open had been picked clean, Dennis advised me to do my pickin' out in the Jungle again. The problem was all the thorny raspberry bushes that lurked among the blueberry bushes. Plus the mosquitoes were starting to come alive. When I announced that I was done, Dennis went into the back and came out with a cold can of beer. We sat and talked, he in the adult swing set, I lying in a heap on the ground. I realized that this was the first time I'd been handed a beer at the end of a hard day's harvest.

"You should've been here last week. In fact the day after you helped us sort, I had a bunch of Laotians come in and pick. They rocked it. Good pickers, too."

"Good pickers?"

"Oh yeah. Clean product. All nice ripe berries; no losers."

"How long did they pick?"

"All day."

"Did they just show up, or do you call them and ask them to come?"

"The guy I sell wholesale to sends them out here. They work in car parts manufacturing. When they lost their jobs a few years ago, they were here all the time, but now that they're working again, they just come out on weekends when he sends them."

"How many pounds per man?"

"About eighty, ninety pounds."

"Anyone do a hundred?" I asked.

"Maybe one or two came close to that."

"Even at a hundred pounds a day, it'd be pretty tough to make a living on that kind of dash," I thought out loud.

"Eh, they pool their money together, those guys."

Dennis sat there smiling at me. What a great smile. And like moths drawn to the flame, various others, some I recognized and some I didn't, appeared out of nowhere to congregate around us, beers in hand, exuding joy and languor. I'd nearly forgotten about this end-of-the-workday feeling what with all the repressive/paranoid/obsessive work I'd been doing over the past twenty years in hospitals, public schools, colleges, and universities. Manual labor is the most natural work of all. No regrets. No dramas. No insecurities. When the day is done, it's done, and it feels so sweet, especially when the weather's nice.

All around me the happy faces beamed with contentment. No one was rushed. No one was harried. No one was thinking beyond the moment. And as I sat there drinking my cold can of beer, listening to them talking about motorcycles and motorboats, skydiving and skin diving, my body and mind purring, I recalled a line from Samuel Butler's underrated classic, *The Way of All Flesh*, in which an eminent London psychiatrist prescribes for his patients suffering from nervous prostration a visit to the zoo with the directions to sit and observe the megafauna, twice weekly.* Were I a psychiatrist or a person suffering from some mental illness, I'd prescribe "Wiltse's blueberry farm, to sort or pick, twice weekly."

Suddenly a voice rang out from back inside the patch:

"Dennis, the cows are in the blueberries!"

"Ahh, let 'em go. I'll put 'em back . . . after I finish my beer."

"Let him finish his beer," Adriano repeated.

On September 1, 2011, I called Dennis in the afternoon and asked him: "Any blueberries left for me to pick?"

"Nagh, they're all done."

"So how did the season shake out?"

"About average. We sold nine thousand quarts. That's about fourteen thousand pounds, plus four thousand quarts wholesale, so about eighteen thousand pounds of blueberries."

"What do you do in the off season?" I asked.

---

* The complete quote reads: "I have found the Zoological Gardens of service to many of my patients. I should prescribe for Mr. Pontifex a course of the larger mammals. Don't let him think he is taking them medicinally, but let him go to their house twice a week for a fortnight, and stay with the hippopotamus, the rhinoceros, and the elephants, till they begin to bore him. I find these beasts do my patients more good than any others. The monkeys are not a wide enough cross; they do not stimulate sufficiently. The larger carnivora are unsympathetic. The reptiles are worse than useless, and the marsupials are not much better. Birds again, except parrots, are not very beneficial; he may look at them now and again, but with the elephants and the pig tribe generally he should mix just now as freely as possible. . . ." Samuel Butler, *The Way of All Flesh* (New York: Dutton, 1946), p. 309.

"Cut hay; chop firewood."

"You sell that?"

"Oh yeah."

"What about next year? Any improvements or additions planned?"

"Well, I'd like to fix the drainage problem. Get the water to drain better and not pool up on the southeast side. Fill in with new bushes where the old ones are burnt out. That's about it. Oh yeah, and I'm getting ready to make my own blueberry wine."

"Really? You going to brand it and sell it?"

"Depends on how it turns out. Hey, if you're down here in the area, stop by for a beer and try out the wine."

"Nothing in the world I'd rather do."

Tomatoes

(Plus potatoes, carrots, beets, parsnips, squash, Swiss chard, lettuce, basil, parsley, blueberries, and peaches)

MANY HANDS ORGANIC FARM

BARRE, MASSACHUSETTS

JULY 2011

At the geographic center of Massachusetts, carved out of the middle of a dark and dreary Hawthornean forest, among lichen-clad hardwoods and moss-carpeted rock ledges, accessible only by dirt roads, is a fifty-acre farm. A plume of smoke perpetually twirls from the central chimney of the homestead, which was built by hand and is made of darkly stained wood. The hinterlands.

Jack Kittredge and Julie Rawson, husband and wife, built this classic, wood-warmed New England homestead twenty-nine years ago. They don't live off the grid, but they are as close as you can get (they need electricity for the freezers and coolers constantly in use). The house faces south with large windows on all three levels, making maximum use of passive solar heat. The old homestead is a straight up-and-down affair, with porches on the top two floors. As you enter from the driveway there is a large mudroom, which opens into an ample living room and kitchen space in the midsection, with one of those old cast-iron stoves that is both heat provider and oven. In front of that is a large kitchen table usually cluttered with freshly picked fruits and vegetables and a large group of people eating the aforementioned. Diagonally across from that is an office on the northwest side of the place. The office is the central nerve center of the Northeast Organic Farming Association: Massachusetts Chapter (NOFA/

Mass),* for which Julie Rawson is the executive director and education director, and Jack is the policy director and editor of the *Natural Farmer.* On all four walls are framed photographs of a range of characters, from family members to NOFA members to farmhands, past and present. The office door is always open and inside is a kaleidoscope of lucubration, such as several desks, all of which are littered with papers and books and computers and lamps. Upstairs there are four bedrooms and one bathroom. The cellar is the seed storage area. A little greenhouse with sliding glass doors is attached to the south side, and is used for seed starting in the early spring

Jack Kittredge is a soft-spoken, Zen-calm soul, around sixty, mustachioed, who reminds me of an English gentleman—David Niven, for example—full of savoir faire but minus the superciliousness. Jack is the strong, silent type. Nods a lot. Will often agree with pronouncements and then add a tidbit of information. Besides writing and editing the *Natural Farmer*, he tends the many fruit trees on the property. And what fruit he's got! Peaches that would make a Georgia peach farmer blush, plums, pears, apples, blackberries, blueberries, raspberries, apricots, mulberries. He doesn't do vegetables, however. That's Julie's bailiwick.

Julie Rawson, Jack's wife, is his foil. Forever in motion, forever on task, forever singing or talking or praising others, she is

---

* NOFA/Mass is a community including farmers, gardeners, landscapers, and consumers working to educate members and the general public about the benefits of local organic systems based on complete cycles, natural materials, and minimal waste for the health of individual beings, communities, and the living planet.

quite simply a human dynamo. She looks just like a fairy godmother out of a Disney version of a Grimm Brothers tale. All she needs are wings—check that: on second thought, they'd probably only slow her down. At fifty-eight and five foot one, she has blond hair like a Scandinavian school girl, and about the same energy capacity. Moreover, her ebullience and insouciance and effervescence are perfectly accented by crystal-blue eyes, which, apropos of everything, are perpetually crossed, or one of them is anyway, giving her the aspect of a blind mystic, or more literally, a cockeyed optimist. When she talks and smiles at you, which is 95 percent of the time, there's a sort of recursive humor effect:* that is to say, as you search the dancing eyes for the center, you get lost in the echolocation of it all. To put it bluntly, being in her physical and intellectual presence is a complete mind and body workout.

I arrived at night in pitch-blackness, pretty freaked out. I had no idea that Massachusetts could be so rural. I stupidly followed the MapQuest directions, which was the wrong thing to do because once I turned off the state route, it was all unmarked dirt roads, straight up and down, with hundred-foot forest walls squeezing me in on either side, and no doubt jam-packed full of weasels and bats that could bite and scratch my eyes out. I don't know

---

* The definition of "recursive humor" is
Recursion
See "Recursion"

A variation on this joke is
Recursion
If you still don't get it, see: "Recursion"

how I eventually found the farm, but luck was on my side as I happened to glance to my right, my head bobbling up and down from the rutted-out road, to see painted on a small flat rock beside a driveway the words "Many Hands Organic Farm." As is my fate in life, the moment I turned into the driveway the car was surrounded by a small pack of dogs. They were barking wildly, but the minute I opened the door, they rushed up to lick my face. Jim, Franny, and Zoe, three happier dogs I have not yet met.

I knocked once and found Jack standing there in the doorway, scratching his balding head, looking at me a little puzzled. It was obvious that my sudden appearance had jogged his memory, helping him recall the fact that I was arriving at the farm that day, and the head scratching was probably a manifestation of his searching his neuron files for the reason why . . . We exchanged a few pleasantries, and within ten minutes he showed me to my room upstairs; told me that breakfast was served at 6:30 a.m. and that work began promptly at 7 a.m.

It was lights out and sweet dreams for me.

The next morning I was awoken by the sound of a beautiful female soprano voice singing an ethereal tune, underlined by the staccato shuffle of prancing feet and an occasional rattling flourish of pots and pans, plates and glasses. The moment my foot hit the kitchen floor from that last step, it was like jumping into a cold mountain stream—Julie Rawson being that cold water:

"Good morning there, stranger. Ho ho ho, you look hungry! Please sit down and eat. Just help yourself." In a little cast-iron frying pan on the table was a vegetable-egg mélange. Beside that was a bowl of peaches swimming in yogurt. I began serving myself and at the same time trying to keep my head above water as the questions and directives flooded in . . .

Where was I from? Could I pass the peaches? What was the name of the book again? Would I mind making some coffee? What was my purpose for writing it? Could I hand her that list of chores at the end of the table? What farms had I visited? Did I ever pull potatoes? Who was my favorite author? Would I mind doing some weeding? How many books had I written? Could I carry those jars of lacto-fermented cabbage over to the counter? What did my wife do? Could I quickly count the number of bags on the chair, thank you? How many kids did I have? Would I mind filling up the wood bin? Where was I going next? Could I quickly sweep the floor? Was I only visiting organic farms? Could I take those bushels on the porch out to the barn? Oh, and by the way, the farmhands coming to work today were all from a prison halfway house . . .

At that moment a half-dozen strapping young bucks and buxom young fillies barged into the homestead without knocking. They were boiling with early morning vitality and youth, nearly equal to that of the blond-haired dynamo spinning furiously beside me. She greeted them all with huge smiles and effusive maternal warmth before letting fly with directives and conditions. It blew my mind to see how it all went down. They stood in line to get their copy of a neatly typed, page-and-a-half-long list of chores to be done; then as they stood there reading and swaying and jerking and mewing hyperactively they listened to Julie's constant stream of verbal addenda. I felt a little weak in the knees. I was definitely in *way* over my head.

They were all just about to race out the door when Julie suddenly called out: "Wait! Wait! Wait! . . ." She introduced me in three quick sentences: "This is Richard. He's writing a book about farming. He'll be helping us for a few days." That was it.

That was all the extra time she could spare. Oh, and: "Ty, would you take Richard with you to help with the animals?"

Then I was sitting next to Tyson, a bear of a young man, who looked like a young Anthony Quinn, early twenties, closely cropped nubs for hair, swarthy complexion, huge brown Mediterranean eyes, barrel-chested, thick forearms, neck and hands. He was as soft-spoken as an anemic librarian, kind of a mumbler, too, but a teddy bear at heart. At each stop and chore, he not only explained quietly what my duties were, but also the reasons why:

Meat chickens — In a shed near the house, behind a large closed door, four score little black-and-gray-mottled-feathered chicks ran frantically, beeping and squeaking and chirping, as Tyson and I filled their water dishes and feed trays . . .

"See that shovel back there?"

"Yeah?"

"Take a shovelful of dirt and throw it in there for them. It's good for their digestion, an's got good bacteria."

"Really? A shovel of dirt?" I dug a nice blade full and heaved it in there. *Dirt for the chickens . . .*

Turkeys — Out behind the pumpkin patch in a grass field are four movable cages, fifteen feet by ten feet by three feet high. They are covered with chicken wire on all sides save the bottom. Inside the cages are dozens of young white turkeys that bob and cluck and chortle and coo—probably a hundred birds all told. On one end of each of the cages are two handles, and at the other end is a dolly. Every day Tyson and another helper move the cages and birds across the field to fresh grass. He took

up position at the handles' end and motioned for me to get the dolly, put it under the frame, and pull.

"Lift it up and move it straight back." In a few seconds we had repositioned the first cage, with the turkeys inside running as we rolled, to fresh grass. They quieted instantly once the cage was set back down, and they began to peck manically at the fresh new greenery.

"The grass feeds them and they feed the soil and the soil feeds the vegetables," I heard him say as he filled their feed trays.

But I didn't need the instruction because it was obvious just to look at it: the large rectangular impressions evolved in ever-greening succession from brown and matted (the more recent days) to lush and green (from weeks past).

"Do the turkeys eventually make their way over to the pumpkin patch?" I asked.

"Oh yeah, by the end of the season they'll make it all over this field and the vegetable beds."

While Tyson filled their feed trays, I washed out their water tubs, which were full of filth, and refilled them. Turkeys are dirty animals and remarkably stupid, too.

Back in the pickup truck: "Yeah, they get sick from drinkin' the water that's full of their own shit. And it's true, they do drown in a rainstorm; that's why we've got the corrugated roof over the entire cage top."

"You ever have problems with weasels?" I asked.

"No, that's what the dogs are for." He moved his chin toward the two dogs running along beside the truck. "But last week I came out here and two turkeys had had their heads bitten off, probably by the coyotes. We just reinforced the cages with more wire so they couldn't fit their heads through."

Pigs — Hidden among a little cluster of trees and boulders at the far end of the farthest field, surrounded by a single electrified wire not but eighteen inches high, lolled the fat pigs. I couldn't believe that they were so unathletic as to not be able to leap the low wire. There were only a half dozen of them, but they were a good two hundred pounds each, shiny and looking almost edible. They were the color of mud-caked camouflage. They huffed and puffed and snorted in excitement as we pulled up alongside. Tyson jumped out of the cab and slung one of the huge fifty-pound bags of feed over his shoulder like it was a down pillow, and motioned for me to do the same with the other bag. Side by side we advanced toward them and their little tumbledown hovel with our sacks of grain. They all raced up to us, their triangular ears flopping like bunny ears as they hopped excitedly once we were inside their space. Their dirt-smeared snouts turned up toward me like they had a mind of their own and nuzzled my leg as I poured the feed into the galvanized hoppers. There were little trapdoors below the receptacles, and they manipulated them easily with their snouts, opening them and thrusting their wet mugs right into the fresh feed like kids bobbing for apples. I glanced around at the landscape and was shocked at how completely trashy it all looked, yet it was nothing but a small island of rocks and trees, perhaps seventy-five feet in circumference. I don't know why I was so surprised: pigs are just like humans, and they have a special knack for transforming sublime landscapes into a welter of filth and contamination—Chernobyl, Fukushima, the U.S. Congress, for example.

Tyson had gone back to the truck and brought out a bucket full of slop—mostly rotten peaches, watermelon rinds, cucumber, and potato skins—and dumped it in a heap next to their

bins. They lunged into it all, oinking ecstatically as they ate, their corkscrew tails quivering with glee.

"No waste on this farm," I thought out loud.

"They love that stuff," he nodded.

"You know, that's what they should be feeding the kids in public school," I opined.

"Yeah." He liked that idea.

Layers — A hundred yards on we pulled up to the layers. Just like the turkeys, they were in movable cages, about the same number of fowl and approximately the same size cages, happily ensconced under the apple trees in the tall green grass. We moved them just like the turkeys, and then set about feeding and watering them and removing the eggs. I noticed that the chickens were separated by species. In two of the four cages were Rhode Island Red chickens, and in the other two were the white Leghorns. I never had realized before that the color of their feathers dictated the color of their eggs, but that's how it looked from where I knelt. The Rhode Island Reds' eggs were all red-shelled, and the white Leghorns' were white. Zoey, a little mixed-breed dog, black and brown with a triangular face that advertised "joy," sidled over to my elbow as I was reaching into the roosts to start removing the eggs. I understood that she was there just in case one of those eggs should slip out of my hand and fall, Humpty Dumpty–style, to the ground.

"Ooops!" There went an egg. I watched Zoe lap up the raw egg in three licks, and then continue licking and licking and licking at the broken shell for a long time.

Cows — Only two cows on the Many Hands Organic Farm. They were close by the chickens, toward the extreme eastern end of the farm, next to the hardwood forest and an ancient stone wall. They lived out in the elements, tethered by rope to a heavy steel bar rammed into the ground with a sledgehammer. All we needed to do was pull up the bars and move them a few feet to an area of new grass and then give them fresh water. They were young cows, and both were steers, that is, they'd been castrated, and what remained of their sacks gave me the willies—six inches of dripping pink skin cinched with twine. They were affectionate bovines, though, and came and nuzzled my hands. They had huge doe eyes with long elegant eyelashes. Their heads were too big, however, and their back ends looked like they belonged to another animal a half-size larger.

"They'll go to the slaughterhouse at the end of the season. Grass-fed beef. Doesn't get any better than that."

"They don't look all that appetizing to me."

"Yeah, but once they're hamburger and sizzling on the grill, I'll bet you wouldn't say that."

He had me there.

Then I happened to notice something odd about the stone wall: from my angle ten yards to the west, there was a section that looked arched, hidden behind the limbs of an overhanging tree. I walked over to investigate it. The three-foot-high wall ran the length of the lower fields, perhaps three hundred yards all told. The entire length of the wall was covered in an ancient white-green lichen, with all sorts of trees growing out and on top of it. When I came to the spot where the arch was, I stood in open-mouthed disbelief. There in front of me, running a hundred feet straight back into the forest, was the longest, widest, highest, and oldest pile of rocks I had ever come across in my entire life. About the size of a tractor trailer, the historic rectangular pile was coated in the same ancient lichen as the stone wall, the stamp of authenticity. Though the stumps of the great trees that had been clear-cut to make way for this oasis in the woods had long since decayed, this amazing monument remained to tell the story of the colonial farmers' untoward struggles and superhuman determination to carve their little slice of paradise out of the wilderness. I remembering reading a book about New England stone walls. The author conjectured that the miles of stone walls in the Northeast at one time stretched farther than the distance to the moon, and that they probably took something like three billion man-hours to construct.*

Brian, the brainy farmhand, came along, and together the three of us harvested peaches for the CSA load that the gang was putting together for distribution later that day. We used these tall, funny-looking aluminum ladders that had an extra-wide base on the front, normal steps to climb up, but only a single pole as a stabilizer on the back. They were easy to position into place among the maze of branches. The peaches were extraordinary. Most of them were the size of baseballs and sweet and juicy as, well, Georgia peaches, *but growing in the woods of central* Mass. Amazingly, Jack grows his fruit trees without spraying any chemicals on them. I was told the chickens spend a couple of weeks under the trees and that makes all the differ-

---

* The book I was referring to was Robert Thorson's *Stone by Stone: The Magnificent History in New England's Stone Walls* (New York: Walker, 2004). Thorson writes, "The stone wall is the key that links the natural history and human history of New England."

ence. And as I moved from treetop to treetop, I noticed something right away—each tree yielded completely different results. The four or five trees we harvested from were loaded with peaches: some of the fruit was ripe and perfect, but the majority was brown and rotted on one side. Perhaps one out of every six peaches we picked was good enough to sell at market. However, there were two trees that were loaded with absolutely flawless fruit. We didn't pick from them, because they weren't ripe yet. In any case, that something so pulpy and sweet bulges and grows out of a woody branch just like that . . . Man! I didn't need them slaughtered, grinded, and grilled to get a hankering for them.

At one point, Brian, who was up on a ladder picking nearby, asked me out of the blue about the book I was writing; specifically, he wanted to know what I was going to say and how I was going to say it. Two tough questions. Maybe it was the fact that I was ten feet up in the air clutching a bucket and reaching for peaches that my answers came out kind of clunky and disjointed. After I had finished, he didn't respond. I wasn't sure how to read his silence. But I was impressed that it wasn't the usual "Are you going to write about me?" kind of question at all; rather, his was a real book reader's question.

With our bushels half full, we loaded them onto the truck and headed back to the barn. On the way back, I asked Brian about the rotten peaches and he explained that the bad fruit would be cut and dried or made into wine or fed to the animals or us.

The barn was a hornet's nest of activity. It was Wednesday, which meant distribution day for half of the 120 CSA members of the Many Hands Organic Farm. Everyone but Jason, the last-

but-not-least employee, was circled around the outdoor sink, washing and drying and boxing produce. Jason had gone to make deliveries.

Once the produce was ready, Julie had a nifty little system set up. With three score multicolored canvas bags, all with laminated name tags—each different color representing a different town—spread out on top of a large folding table, we walked down the line with our assigned produce bundles and placed them in the bags. There was lots of banter and silliness along the way. Two "trustees," as they referred to them, from the CSA were there helping pack and load. I learned that CSA trustees who help to harvest and pack once a week get free shares during the season.

It was quite a satisfying feeling when all was said and done to see that lineup of bright-colored bags bulging with vegetables and fruit and ready for the fridge. I helped one of the trustees load a couple dozen orange bags into her van, and then said goodbye to her as she drove off.

Lunch was served al fresco on a yellow picnic table under a peach tree, in large stainless steel bowls and cast-iron frying pans. What a feast! Heaping mounds of fresh vegetables, cut fruit, green salad, and sausage and onions, and plenty of cold water. It was the same deal as the Tantré Farm, where the helpers prepare the meals. Nicole had done the cooking and fixing that day. At the far end of the picnic table sat Jason. He charmed the hell out of me with his off-color comments and thick New England accent. It's a bit of a stretch, but squint your eyes and think of a shirtless, broad and burly, working-class version of

Groucho Marx, with the kinky black hair, the black-rimmed glasses, the quick wit, and a baseball cap turned backward on his head. Oh, yeah, and every other word out of his mouth was absolutely filthy:

"Fahkin' Jules, you fahkin' did it to me again! You dumped a fahkin' shitload of hot spice in the Swiss chahd, didn't you? You know I fahkin' hate when you do that." He punched at the air, then grabbed his glass of water and downed it in one gulp; poured himself another and downed it just as fast.

He was talking to Julie, who sat in the middle of the table watching over all of us with her eyes bobbling like a mother goose . . . and eating like a truck driver.

"Come on Jay, you know I wouldn't do that to you," she said in her angelic little voice, giggling merrily, with a bulging mouthful of sausage.

Inasmuch as his language was offensive, it was nonetheless dripping with reverence and profound adoration for his employer. And that Julie is the type of person who has cursed maybe three times total in her entire life made Jason's playful comments seem even that much more laced with love:

"Yawr a fahkin' evil witch, you know that? I'll bet yawr fahkin kids were scared shitless of you. You probably washed their fahkin' mouths out with fahkin' jalapeño peppers or fahkin' Tabasco sauce."

She giggled and giggled as she continued to pack away her chow, pixie dust sprinkling from the salt shaker.

Next to Jason sat Brian, shaved head, piercing blue eyes, wiry build, just a hint of a New England accent. He was the more serious and intellectual of the group, and every time Brian spoke, people quieted and listened. Across from those two were

the two ladies, Margie and Nicole. They were all about the same age, in their early- to mid-twenties, and their smiling faces glowed with youthful beauty. You could tell by the way they sat there in their sleeveless shirts and straw hats and bare feet and glowing cheeks watching and listening that they loved every moment of being down on the farm. And they did not suffer Jason's off-color barbs gladly. They often jabbed back, good and hard. Tyson and I sat at the far end of the picnic table, both of us quiet as monks, but racing to see who could shovel his lunch down the hatch faster. He won easily. The one remaining CSA trustee who had helped make the lunch had long since wandered back up to the house.

After lunch I helped Tyson drench the tomato plants because they were looking pretty sad. Drenching is an interesting subject that I was not aware of until then. There was a funny moment when I asked what we were doing next. Tyson was driving and facing forward. I've got a bad left ear and he's a bit of a mumbler. Here's what I heard:

"Dredging."

"Dredging?"

"No, dressing."

"Dressing?"

"No! Drenching."

"Drenching?"

"Yeah."

"What's that?"

He parked the truck next to the tomato rows and went around to the back and pulled out two large watering cans and a half-gallon jar with the word "Foundation" written on it. It looked like chocolate syrup.

"What's that stuff?

"It's the drench."

"I figured that. What's in it?"

"What's in it?"

"Yeah."

He poured about a cup into each of the two watering cans and took the hose and began to fill them with water.

"You want to know what's in this stuff?" He took a breath: "Nitrogen, phosphorus, potassium, chlorine, calcium, sulfur, magnesium, iron, manganese, boron, zinc, copper, and molybdenum. Oh, and selenium." Then he explained: "There are nutrients that the vegetables need. Some of the nutrients they need in large concentrations and some they need in smaller concentrations. The macro- and the micronutrients. That's what this stuff is, all mixed together."

"Hmmm, sounds like science to me."

We then drenched the eight rows of tomatoes.

After drenching we went around and checked in on the animals again. The day just sort of faded away, and before I knew it Jack, Julie, and I were crammed three in a row in the cab of the pickup truck, heading for downtown Barre, where we were to meet the director of the Quabbin Community Band. Julie was the French horn player in the band, and also in charge of getting the equipment to the gig. She was dressed in the traditional white shirt and black pants, and as a former community band member myself I almost wished I had brought along my trumpet.

Once the director arrived, we went down into the basement of the gazebo and hauled up the music stands and the two large timpani, put them all in the back of the truck, exchanged platitudes, and raced toward Princeton, Massachusetts. Julie rode with the director while Jack and I rode together in the truck. It was the perfect opportunity to get to know Jack a little better and to ask as many pointed questions as I could:

"So what do you guys do for health insurance?"

"We put money aside each month, and if we need to go to the doctor or hospital we just pay for it."

"Whoa, really?"

"For example, I had an operation last year, and so did Julie, and we paid for it in cash."

"That's got to be prohibitively expensive?"

"You shop around a little bit. But we can afford it. We built our house ourselves, so we have no mortgage. We heat it with wood and have no heat bills. We grow and raise our own food, healthy food, without all the chemicals or toxins that make you sick."

"Then the farm makes enough money so that you can save for your retirement?"

"No no, we don't make our money from the farm. Julie and I both work for NOFA, and that's where the bulk of our income comes from. I know that Julie would eventually like to get the farm to the point where we're making some money, but we're not there yet."

I was in awe of Jim and Julie—how they are able to simply remove so many of the present-day balls and chains that keep the rest of us bound and beholden to humiliating jobs and mean-spirited bosses and inhumane working conditions. These independent, hardworking, creative, and competent homesteaders aren't going to be held hostage by health insurance companies or banks or corporate fearmongering or anyone or anything, for that matter. They have found their own way to live a freer and better American lifestyle. What an inspiration!

I had another burning question:

"So how old is the farm anyway?"

"How old?"

"Yeah, I saw a pile of rocks today that must go back to when they first carved the farms out of the wilderness back around the time of King Philip's War, when the Indians nearly ran the colonists off the continent?"

"King Philip's War, when was that again?"

"Late 1600s."

"I don't know about that, but definitely the late 1700s. This area and the old farms around here were all involved in Shays' Rebellion."*

"Cool" was all I could think to say.

Then as if by magic, I was standing on a high hill, in the middle of a classic New England village green at the long end of a summer's day. Princeton, Massachusetts, is something out of a dreamscape. With white churches and Federal-style country

---

* Daniel Shays was a simple farmhand from Massachusetts at the start of the American Revolution. He fought in some of the most important battles of the revolution, including Lexington, Bunker Hill, and Saratoga. Wounded and spent, he resigned from the army. When he returned home in 1780, he was hauled before the court for unpaid debts. It didn't take him long to realize that he was not the only one who was being hounded for nonpayment of debt accumulated during the War for Independence. He organized a rebellion, and the rest is history. See Howard Zinn, A *People's History of the United States: 1492–Present* (New York: HarperCollins, 2005), p. 93.

mansions on all four sides of the square, a dreamy blue sky above, and a 200-degree view of the surrounding area, it was positively breathtaking. And from where I stood, just west of the bandstand at the eastern edge of the green, I could actually see Boston! It was a thrilling sight for me to look down across the horizon and pick up the shadowy outlines of the Prudential and Hancock buildings, my old college stomping grounds, in the ionized, blue-gray distance, fifty miles as the crow flies. Then the music swelled, and I lay back on the soft grass, staring up at the infinite sky. After a nine-hour day working down on the farm, it was all I could do to keep from falling asleep.

It was a beautiful night, albeit dark, and as we drove back through the overhanging forest along the winding roads, toward Barre, three across in the cab again, I took the opportunity to ask more questions, this time about the halfway house gang. Julie did most of the talking.

"So tell me again how you got these halfway house folks?" I asked.

Julie filled in all the details without hesitation: "We got a call from the director of Dismas House and the Almost Home program at the Worcester County House of Correction in March of 2007. He wanted to bring out a van full of guys once a week to work on the farm, strictly as volunteers. The first day was April 20, 2007. Ten or so guys would come out each week and do manly-man things like running the rototillers, chopping wood, building trellises, cleaning out the chicken houses—whatever we needed. They stayed for lunch and then returned to the Almost Home program. Brian stood out early as a really hard worker, and when he graduated the program, we hired him to work full-time. But we had to up production to pay for him. And we got to where we are today. Jason was one of the last participants from Almost Home before the project got defunded and shut down. Tyson was at Hope House in Leicester—where we, through the Many Hands Sustainability Center, picked up guys each week. This was last year. And Nicole is from the Linda Fay Griffin House in Worcester, where we picked up women this year. Brian, Jason, Tyson, and Nicole are on staff. A few other guys have come and gone on staff since 2007."

"So you said you upped production on the farm to pay for the help. How much did you up it?"

"Since then? A hundred percent."

"A hundred percent!"

"It's the people."

"They're the best we've ever had," Jack added.

"Yes," Julie continued. "Just look at a person like Brian, who's so smart, so full of different ideas and energy and vitality, but you put him in an educational system that just wants him to conform and memorize and do what they tell him to do . . . he couldn't do it. So what were his options? Getting in trouble, doing drugs, that's about it, but—"

"But the farm, the animals, the tractor, the fruit trees, the cutting wood, the fresh air, the independence . . . it's a perfect place for someone like Brian." I couldn't help but finish her thought.

"Yes. Yes. I know. I know. There's something about working on the land, the connection with the soil and the plants and the animals . . . That's where recovery is found. This might sound corny, but I don't feel like I'm farming vegetables or fruit or meat animals anymore. My main crop is people. Now that these folks have joined me, I'm farming people. That's my primary product."

It was a very profound statement that Julie had just made, and as she continued to tell me more about the backgrounds of each member of the halfway house gang, in loving terms, a thought began to grow inside me . . .

Yes. Yes! Yes!!! The answer to so many of America's social and economic problems is obvious—agriculture! Agri-culture, the first word in culture. Like the scene from The March of the Wooden Soldiers *when Stan and Ollie are searching around the toy shop for the last remaining darts to shoot at the invading Bogeymen when they suddenly look around and realize that they have one hundred strapping, six-foot-tall wooden soldiers at their disposal to do the fighting for them. With over two million people languishing behind bars in the United States, one of every one hundred adults, we have an unused workforce on a scale that defies credence. The farms are here; what's needed are the helpers, the farmhands, the Brians and Tysons and Jasons and Nicoles. We need to turn our prisons into farms, organic farms, and all of the inmates into stewards of the land. A million people, in the prime of their lives, tilling and sowing and harvesting this nation's crops . . . that is, if rehabilitation is our mission . . .* And that's when that all-nasty conspiracy against the obvious farted in my face (oddly, in Jason's voice and words): *"Yawr a wicked fahkin' genius, ahn't you! Do you know how much fahkin' money they fahkin' make off of prisons, Scribble Boy?"*

The next day I got my ass kicked. Nicole and Jason and Margie didn't show up to work for some reason, and Julie had paper-work to do, so I was put under Claire's charge, or perhaps Claire assumed charge of me . . . Whatever, Claire immediately got under my skin. Tall, athletic, with a model's physique, Claire, who was about thirty, was easy to look at but hard to take. I thought at first that Julie had jokingly tasked Claire with putting me through my paces, as she ordered me about, waving her finger and squinting with annoyance, telling me to do this and to do that and not to do this or that, and to do this and that this way, not that way, but then I noticed that she was equally as exigent with Brian and Tyson and the two CSA women. Funny thing was, they didn't seem the least bit annoyed by her. Just me.

At one point she handed me a hoe and instructed, via a quick demonstration, how I should carve a trench for the planting of kale. So I began my work, taking extra care to make sure that the trench, my trench, was as straight and as equal in depth as I possibly could make it. A few minutes later, I looked up and there was Claire hoeing down the row, *my unfinished row*, directly at me, going three times as fast as I was. I had to literally jump out of her way as she haughtily whizzed by and did over my work. It was the cutting edge of passive-aggressive behavior.* At that point I became aware of how she had managed to excavate my adolescent Rolodex of He-man Woman Hater's Club slogans, as I began growling under my breath: "Little Miss Bossypants." "Goddamn women take all the fun out of work." "Agh, your mother probably *does* wear army boots."

In retrospect it was a funny situation for me. Here I had been a teacher for nearly twenty years, but apparently, down on the farm with my hands in the soil, my emotional default position is

* Pun intended.

that of a ten-year-old. I was ashamed. After all, as a teacher for decades, I had instructed scores of type-A Claires. It had been my job—my professional duty, in fact—to teach these pretty little perfectionists that when they are working together in a group, they must take care to be patient and encouraging and respectful of others' feelings, abilities, and work ethic. In any case, for the rest of the day, I put my game face on, bit my lip, and said nothing—but continually reassured myself that had she been in my classroom, after a stunt like that with the hoe, *why I would have . . . I would have . . . told on her . . .*

That night at dinner, Claire was the first topic of conversation:

"So what's the deal with Claire?" I put it to Julie flat out.

"Deal? Claire?"

"Yeah, did you put her up to that today?"

"Put her up to what?"

"Busting my chops?"

"Nooo." Julie seemed very surprised. Realizing that I was perturbed, she immediately put her fork down. She sat there with an ever-broadening smile, her cross-eyes dancing wildly, relishing this tattling session. (Jack was out of town for the night, and sadly not able to weigh in on the conversation.) I continued: "Well, it was all I could do to keep from laying into her, I can tell you that."

"Really? Was she mean or disrespectful to you?"

"No, she wasn't mean. But she was really bossy. I . . . well . . . I . . ." I was sputtering for words, so I just launched into the hoeing-my-row incident, my heart pounding in my chest . . .

"I'm going to have to have a little talk with Claire about all of this."

"I wish you would." I salted and peppered.

After eating in silence for a few minutes, I changed the subject and asked Julie how it was possible for her to double her output in just five years. I assumed her income had doubled, too. "So how much do you make in a year from the farm?"

"Gross?"

"Sure, gross."

"A hundred thousand."

"A hundred thousand dollars?"

"That's gross. Net is a different story. Net is like negative fifteen hundred."

"But still . . ."

"Yes. I consider that a success."

"And what do you attribute your success to?" I asked.

"The right leadership, the right workers, and the fertility of the soil. The animals add the nutrients and the plants just keep growing bigger and better, and all we have to do is plant and pick."

"The chickens and the turkeys and the cows and the pigs and the dogs and the drenches?"

"And the right people."

"That's the magic formula?"

"For the Many Hands Organic Farm it is."

The next day was Friday, my last day on the farm, but another wild CSA day. The halfway house gang had returned in full, and fortunately for me, no Claire. About midmorning I found myself on hands and knees in the potato patch, digging potatoes with my bare hands, just like everyone else. In fact the entire cast was

in the potato bed, digging around. Jason was at the far end with Brian, pulling weeds so that the rest of us could find the withered stalks of the potato plants and have an approximation of where the potatoes might be located. Tyson and Nicole and I were at the back end of the line, elbows deep in the loose soil, and Julie and Margie and one of the CSA trustees were following closely behind the two weeders.

The banter was lively, with Jason punctuating often with his expletives. The trustee, a nice woman around forty, was talking about her first-ever flock of meat chickens, and the fact that she had to slaughter them herself. She wasn't looking forward to it, nor did she know how to do it. Julie made the comment that she thought drowning them was the easiest and most humane way. Tyson said he just broke their necks. Julie disagreed with Tyson. I piped up and told her how my wife's step-grandmother, who used to raise chickens for a living, would cut open the aorta vein in the back of the throat with a razor blade and bleed them to death. Jason pooh-poohed all of us and said: "I just fahkin' smash theyh heads with a brick." But none of the suggestions was making the poor woman feel any better.

"I don't know if I can do it," she sighed, full of despair as she felt around blindly for more potatoes in the dirt. "The thought of killing them just . . . just killing them . . ."

"Well, remember one thing," I offered sincerely. "When the Navy SEALs broke into Osama bin Laden's lair in Pakistan, you know who was there guarding him, don't you?"

She had no idea. She stopped her digging and looked over at me with a hopeful expression.

"Chickens. They were all circled around him, smoking those foreign cigarettes and pecking holes in an American flag on the floor. They hate us, you know. They hate everything about us, everything we stand for: our freedoms, our way of life, our celebrity chefs—Kenny Rogers, Roy Rogers, Colonel Sanders . . ."

She started laughing good and loud.

"You're fitting right in, Rich." Julie looked over at me, beaming with approval. "The best thing about working here on the farm with this great group of people is that we get to have these kinds of conversations and laughs all the time. It's wonderful."

"That's fahkin' bullshit, Jules. The best thing about the fahm is that I get to steal Tyson's cigarettes," You-know-who joked.

"So that's where my cigarettes—! Damn it, Jason! I'm saving for my license!" Tyson growled.

Just before the CSA push, we all went over to the tomato patch to pick tomatoes. The sun was high in the sky, and the humidity had crept in. I was sweating bullets from all the bending and crawling through the tall, close rows. Brian had instructed me to remove the little leaf caps from the tops of the tomatoes, and that forced me to take each individual tomato before I dropped it into my bucket and pluck the little cap off with the tips of my fingers. It was like removing lint off a shirt, especially with the tiny cherry tomatoes. And it was while I was delicately working off a cap that a big tomato went whizzing past my head. Within moments, I was smack in the middle of an old-fashioned tomato-hurling fight. It was Nicole and Margie and Brian and Tyson

against—surprise, surprise—Jason, who was hopelessly trying to defend himself. Julie and the CSA trustee were back at the barn preparing for the bagging, which might have precipitated the event. I just crouched down below the tomato cages and watched Nicole and Margie, who were nearby hurling rotten tomatoes down at Jason, with good success.

"You fahkin' bitches! You think I won't fahkin' [*splat*] get [*splat*] fahkin' get even with youse. Wait till I fahkin' [*splat*] fahkin' [*splat, splat, splat*] . . . grrrr, yawr gonna fahkin' regret this!" *Splat! Splat! Splat! Splat! Splat!*[*]

Standing there in the kitchen, saying goodbye to all of them, Jack included, it was suddenly so obvious to me that the Many Hands Organic Farm was like the real-life, happy-ending version of *Animal Farm*. In the utopian concept, the underdogs take control of their fate, and together work for the good of themselves, society, future generations, and the land itself. As I looked at them looking at me with huge smiles, it was overwhelmingly obvious—it doesn't take nations, ideologies, revolutions, pogroms, coups d'état, totalitarian regimes, gulags, propaganda machines, secret police, death squads, assassinations, torture, cults of personality, brainwashing, prisons, atom bombs, Great Leaps Forward, espionage, censorship, university administrators, or martyrdom to create better working conditions, safer neighborhoods, healthier environments, in sum—*civilization*. All it takes is a handful of people with love and

* I don't think anyone saw me, but I threw one, too.

understanding, a dozen acres of lush tilth, a sense of responsibility, and a desire to lead with compassion. Funny, but I don't ever remember learning that in grade school, middle school, high school, college, graduate school, or even while teaching at the university . . .

Red Raspberries,
Brussels Sprouts

ANN'S RASPBERRY FARM AND SPECIALTY CROP

FREDERICKTOWN, OHIO

EARLY SEPTEMBER 2011

Ohio is a nation state. Bordered by the Great Lakes to the north, the mighty Ohio River to the south, with hardscrabble cities from one side to the other—Youngstown, Akron, Cleveland, Toledo, Columbus, Dayton, Cincinnati—it is the birthplace of the Wright Brothers, Standard Oil, and more U.S. presidents than any other state.* Home to the perennial powerhouse Ohio State Buckeyes as well as the largest Amish population in the world. *O-HI-O!*

I went to college there for two years, or I should say, I went to play football there, but in the end neither worked out, and I fled back east to Boston, where I belonged. Ohio was an utterly foreign nation to me, full of cows and trucks, cornfields and silos, farmers and factory workers. Oh sure, there were plenty of pretty coeds and beer to distract me, and ivy walls and ivory towers to protect me, but I could still feel what was going on out there: *Work*, with a capital *W*. Not the moneygrubbing, ladder-climbing, make-a-killing-while-you-can kind of work like the men in suits with briefcases did back east, but the nose-to-the-grindstone,

---

* William Henry Harrison, Ulysses S. Grant, Rutherford B. Hayes, James A. Garfield, Benjamin Harrison, William McKinley, William Howard Taft, and Warren G. Harding.

get-your-fingers-dirty, one-foot-in-front-of-the-other kind of work. They even played football that way. No, Ohio didn't make any sense to me way back then.

But thirty years later it made perfect sense as I turned off the interstate east of Columbus and drove right into the belly of the beast, with the pig shit ripe, and the cornstalks sky-high, and the pickup trucks lined three deep in the driveways, and the grown-old down linemen ambling out to the barn in their boots and overalls and John Deere hats. And deeper in, the pickup trucks gave way to black buggies and horses, and the grown-old linemen in boots and overalls transmogrified into bearded elders in plain outfits and wide-brim straw hats. I was on my way to pick raspberries right in the middle of Ohio.

It's a great leap of faith to invite someone you've never met to stay in your home for three days and nights. It's an Olympic long-jump record to allow that person to share a bathroom with your teenage daughter. And it's an Evel Knievel Snake River Canyon ride to expose your whole family history, your wife's included, to a writer. But there are people out there who leap through life because they just can't walk.

Ann's Raspberry Farm is located in the middle of Amish country, squeezed between copious acres of Amish land and a huge conventional soybean farm. The simple name captures the essence of the whole shebang—the people as well as "the farm." The house, for example, looks like a standard suburban model—two stories, two-car garage, front porch, back deck, vinyl siding, neat and tidy. There's a wide patch of brussels sprouts followed by a U-pick raspberry patch with about twenty-five rows of bushes, each row about fifteen yards long, on the west side of the driveway. If you continue up that side of the drive, there is a horse barn with a white horse in it, and a hoop house full of raspberry bushes. On the other side of the driveway is another quarter-acre patch of brussels sprouts in the front part of the yard and a regular green lawn for the rest of it. Out in the backyard, along the Amish hay field, is a small garden full of jalapeño pepper plants and a few tomatoes. All told, about five acres. That's the entirety of the farm.

Daniel and Ann Trudel moved to their Fredericktown house eight years ago. Neither had any background in farming. Daniel was a market researcher, a demographer. Still is. He grew up in Montreal and didn't learn to speak English until after he was married, at age twenty-six. Ann was a piano teacher. Still is. They met during college, down in Florida while both were on spring break. He transferred from McGill University to the University of Akron so he could be near Ann. After marriage, Daniel took a job in Michigan at a marketing firm, and they moved there and lived for fifteen years, producing two children, a boy, Eric, and a girl, Allison. But then a great family trauma literally forced them into hiding. So Ann's closest childhood friend, Betsy Bradley, who lives two doors down from them now, invited them to stay with her and her family. They came and stayed awhile, and it just so happens that the house that they now live in was for sale at the time. So they bought it. Daniel was able to keep his job and work from home. And it was in that way that they were able to slide into their new lifestyle, and out of harm's way. As for the farming . . .

It's no exaggeration to say that Daniel bounded out of his

garage to greet me as if he'd been shot out of a cannon. With his straight brown hair and blue eyes and French accent, he's a dead ringer for a fifty-year-old Yves Montand.* One of the most pleasant, bursting-with-enthusiasm, and totally over-the-top people I'd ever met in my entire life, he's like the amiable version of the Tasmanian Devil, only instead of drooling and snarling and holding his curled hands as if he's ready to tear you limb from limb, he smiles and nods and makes movements with his hands as if he's clutching two specimen jars that constantly need agitating. Within the first half hour, he'd so overwhelmed me with his positive energy and use of superlatives and hyperbole and nonstop chatter that I got to feeling nauseous. It was too much, too suddenly; I'd been driving a long time in total isolated silence.

Daniel took me directly into his hoop house, where the Trudels grow the red raspberries for their award-winning Jalapeño Raspberry Jam.† It was a medium-sized greenhouse, perhaps ninety feet long by thirty feet wide. He introduced me to nearly every single raspberry bush in the place. In my entire life I'd never seen anyone so proud of a bunch of spindly plants. With his head bent, his eyes nearly popping out of his head, he'd grab hold of one of the ripe little monsters hanging down

---

* *Manon of the Spring, Jean de Florette, Grand Prix.*

† Ann's Raspberry Farm and Specialty Crops won first place for its entries of Savory Sprout Relish in the pickle category and Jalapeño Raspberry Jam in the preserves category at the 2010 Good Food Awards competition. It is an amazing accomplishment for new farmers to win awards alongside professional artisanal jam makers and picklers. They won the same awards in 2011 and 2012.

in front of his face and exclaim: "Isn't that *amazing*? Huh? Isn't it? *Isn't it!?*"

Yes, the raspberries were huge, red, and bulbous as clown noses, bending the ends of their branches down like weights on the ends of fishing poles. And there were so many of them, too, that if you didn't have a good grip on reality (which I didn't—*why else would I mention it*?), you'd swear that you were seeing red dots everywhere. Daniel showed me how he had originally tied the bushes up with wire so they could be held in place at six feet tall, but recently he had had to add two more feet of wire because the plants were *growing right up through the roof*. ("Isn't that *amazing*? Huh? Isn't it? *Isn't it!?*") And just when I thought I'd have to jump in my car and speed off in retreat, he put me to work, by myself, picking the raspberries, but not before issuing me both a challenge and a promise:

"If you pick all of the ripe berries in this room, in two days there'll be even more, I guarantee it."

He set me up with a little wooden carry-around stand on which to haul my load of green quart cartons. The knee-high carry-around had a long handle like one found on an old wooden toolbox, and held eight quarts. It was the perfect chill-out after an extreme half hour with Daniel. Within minutes I was completely in the zone. The sensual feel of those pulpy, hirsute dollops was soothing my savage breast. I was truly making love to those sexy little berries. Thrilling how they would give themselves over to me, dislodging from their triangular moorings at my slightest urging. And it was while harvesting that lascivious fruit that I realized how berry picking had become for me a new type of meditation; a physical act of poetry; a silent performance of music;

a nonviolent form of shopping.* Berry picking by oneself exalts the mind and adorns the human spirit in pearls and décolletage.

I have no idea how long it took me to pick every single ripe berry in the place—hours, perhaps—but when I was done, I was absolutely, one hundred percent certain that there was not a single ripe berry left in the house. Proud as a peacock, I marched out of there with my heaping quarts of berries crammed full on the carry-around. Daniel was in the garage, marshaling his goods and wares, and when he saw the berries, I thought he was going to kneel down and kiss my hand.

"It's that *amazing*? Huh? Isn't it? *Isn't it!?*"

And then—*boom*—he had me in his van and we were driving over to the Amish neighbors to pick up some sticky buns for breakfast. Again, his energy level and effusiveness had me feeling slightly queasy almost the moment we were out of the driveway. I was worried at times that he was looking over at me too often instead of watching the winding, narrow road in front of him. He was giving me a quick lesson about the Amish. He had firsthand knowledge and spoke glowingly of them, using words like *talented, reliable, resourceful, dedicated, generous, reasonable.* The latter word piqued my interest.

"Reasonable, you mean like agreeable and rational?" I asked.

---

* As I write this, a story making the rounds on all the major networks reads: "LOS ANGELES (AP)—A woman trying to improve her chance to buy cheap electronics at a Walmart in a wealthy suburb spewed pepper spray on a crowd of shoppers and 20 people suffered minor injuries, police said Friday. . . . The attack took place about 10:20 p.m. Thursday shortly after doors opened for the sale at the Walmart in Porter Ranch in the San Fernando Valley."

"No, I mean like cheap. I have them do work for me all the time. They charge half what the regular contractors or companies charge. They can build anything."

"Are they good neighbors?" I pressed.

"Sure they are. If you ask them to do you a favor, they consider it an honor. If you do them a favor, then they feel obligated to you. Actually, they don't like that. One time they asked me to drive them to Columbus for a doctor's visit. I drove them and then refused to take money for it. They didn't know how to handle that. So you know what they did? They came to my door the next morning with pies and bread and a box of fresh vegetables."

Up ahead a large farmhouse loomed to the left side of the road, with silos and barns and a swing set all around it, like a picture out of *National Geographic*.

It was a classic Amish house, square and sturdy, with a front porch and two large barns, one behind and one to the side. A young girl in a long blue dress, an apron, and a white bonnet was hanging laundry on the clothesline just outside the screened-in kitchen door. There were two teenage boys in plain dress—black pants, blue shirts, suspenders, straw hats—wiping down a buggy outside the horse barn directly in front of us. An elder was behind them working on a plowshare, same clothes as the boys' but with his sleeves rolled up. A flea-infested German shepherd lay on the ground off to the side of the van, scratching itself maniacally and paying us no mind. Cats of all colors and stripes littered the yard. Daniel slowed, then jammed the van in park and jumped out and skipped boldly up the walk toward the house. He stopped and gave a wave and a shout out to the elder and the boys, said something to the girl hanging laundry, and then walked in the screen door without even knocking. I was amazed to see him reappear moments later holding a bag of goodies and talking back over his shoulder as he bounded toward me in the van. Never a wasted moment or movement with Daniel.

When we got back to the house, Ann was in the kitchen making a quiche for the next day. She was over by the stove, giggling at us as we came in. She reminded me of Paula Deen, only about twenty years younger and without the fake-white teeth. Ann has a sweet, mellifluous soprano voice, and a personality to match. Very calm and composed—Daniel's perfect complement. She was looking sympathetically at me with her big brown eyes. I knew what that was all about before she even said anything: "Has he overwhelmed you yet?" she chuckled.

"Oh yeah," I sang.

It's so refreshing to meet people who understand themselves and each other.

Ann showed me to my room; it was downstairs in the basement—a large queen-size bed, firm, just the way I like it, and very private. They were the best accommodations yet.

Ann had a homeschool event to go to that night, so Daniel was taking me into town to the thirty-fifth annual Fredericktown "Tomato Show," where we would hopefully find something to eat. I told Ann that she and Daniel were keeping my perfect record alive.

"What record is that?" she asked.

"Homeschooling. Four in a row."

"Really? So the other farms you visited homeschooled, too?"

"Yes. I am finding a lot of recurring themes . . ."

"Really, can you give us some examples?"

"How about this one: long-lasting, strong marriages with exemplary teamwork."

She seemed pleasantly surprised. "Wow!"

Daniel wasn't saying anything; rather he was involved in making a list of chores for the next day. Another common theme . . .

In the time it took us to drive to town, Daniel revealed to me some pretty extraordinary facts about Ann's Raspberry Farm. First, he told me that they had doubled their sales over the past year. That included their value-added products—the jams and relishes—plus the farm products themselves—the brussels sprouts and raspberries that they sold via the farmers' markets and U-pick enterprise. He explained the concept of the value-added product to me—basically, additional goods or services. Like an apple orchard that sells not only apples by the bushel, but also apple pies by the tin. As of that day, he claimed, they had picked some seven hundred pounds of raspberries even after the spider mites had ravaged the bushes and nearly wiped them out earlier in the season. And it was the very first time since they started the farm seven years ago that they had to invest in labor, and it paid off in spades.

"So you and Ann and Allison and Eric did all the work, all the picking and prepping and jam and relish making prior to this year?"

"That's right."

"And so just by hiring labor that doubled your sales?" I interpreted.

"No, that's not it. When we hit our maximum capacity the year before, we realized two things: that we would need to get labor, but the key was that we also needed to resist the temptation to increase production, which is what most businesses would do. Instead we took the advice of the father of organic farming, Eliot Coleman, and increased the price of our products. And that's how we doubled our sales."*

"You upped the price on all your products?"

"Yes."

"And so now do you make your living exclusively from your farm?"

"Not yet. But if we double again, we will."

The Tomato Show was iconically American Lame. Fredericktown has an old, Appalachian downtown feel, with classic stone and brick nineteenth-century buildings lined up on either side of the sloped main street. There was a large stage, all lit up, set in the middle of the main thoroughfare with an enormous crowd jammed cheek by jowl in front of it. There was a talent competition under way at center stage, but all I saw was lip-synching, my pet peeve.† The crowd was nonetheless more than generous with its support, whistling and clapping appreciatively as each new participant gave it his or her all.

We toured the entirety of the fair: the food booths, the raffle stands, the shooting galleries, and the big-top tent, under which the winning vegetable entries were on display. Other than by a

---

* *The New Organic Grower: A Master's Manual of Tools and Techniques for the Home Market* (Chelsea, VT: Chelsea Green, 1995) is Coleman's most famous work.

† As a former talent show director at more than one high school, I take umbrage with lip-synching, having banned it from any competition that I ran. How is lip-synching a talent?

few of the silly faces painted on the various gourds and squashes, I was not in the least bit impressed.

After a half hour, we found ourselves seated on top of a picnic table at the end of the street, eating sausage sandwiches.

I had to ask: "So what's the ultimate tomato tribute?"

"What do you mean?" he asked, chewing heartily.

"Is there a huge cash prize for the biggest or tastiest tomato or spaghetti sauce?"

"No, not that I know of."

"Isn't this some kind of tomato harvest celebration?"

"Not really," he said with a mouthful of food.

"So then why do they call it the Tomato Show?"

He stopped chewing and looked at me with those huge, intense eyes of his. "*Je ne sais pas*! But I'm going to find out."

He was up and chasing after an answer in three shakes of a stick, looking for an authority figure to answer the $64,000 question. Finally, up in front of the town hall, he found the right person. The man was delighted by the question, too.

"You want to know why we call it the Tomato Show?" the short, elderly man dressed in a red T-shirt with the Tomato Show logo on it guffawed, his expression full of italics and exclamation marks: "I'll tell you why: because they wanted to put on a town fair, but they couldn't agree on a name. No matter how many names were suggested, they couldn't agree on one. *Not a one!* Finally, when it was either they decide on a name or forget about the whole thing, someone just blurted out, 'Tomato Show,' and no one disagreed with it, so *that's* why we call it the Tomato Show."

"There's no connection at all to tomatoes?" I clucked.

"None whatsoever."

"Someone blurted out 'Tomato Show' and that was it?"

"That's right."

"Like a show about nothing, a town fair about nothing," I said sort of under my breath. He shook his head in agreement while I scratched mine.

In Europe they have grand festivals and celebrations commemorating preternatural apparitions of the Virgin Mary, miracles performed by patron saints, legendary and historical events and figures, and one-of-a-kind cuisine. Only in America could some chucklehead shout out "Tomato Show" and *wham bam*, that's good enough to close off Main Street, set up some cheesy carnival rides and funnel-cake booths, erect tents, judge a few baked goods and jarred vegetables, and hold a lip-synching competition. We are, by God, the most absurd but spontaneous nation of all times.

Somewhere around midnight, I stood on the Trudels' back porch, alone, looking up at the twinkling stars, the black night like a close friend next to me. And off in the distance was the most enchanting sound I'd heard in a long time—the clip-clop of a horse-drawn carriage. The Amish. The sound didn't seem to wax or wane, but continued in one suspended refrain, like a continuous note. The exotic sound consumed my every thought. At one point, just before the clip-clop started to fade, I could see the winking glow of a lantern on a carriage, like the light of a boat out in the water, a half mile off.

The next morning I was awakened by Daniel running up and down the cellar stairs several times. I don't believe he was doing it for my benefit or for the exercise; he was just everywhere at once—hopping, skipping, and jumping through his daily chores. I crawled out of bed, flopped around on the floor in an attempt to do some stretching exercises, and then got dressed. Up in the kitchen on the counter were the Amish sticky buns, a plate, a pot of coffee, a cup, and a little pitcher of milk waiting for me. I stood and ate and drank. It didn't take long before Daniel came careening into the kitchen, his motor revving. He idled down, and stood with me, chatting.

"Feel like picking brussels sprouts today?"

"Sure."

"You ever pick them before?"

"No, never."

"Piece *awf* cake." His French accent was pronounced this morning.

"About how many bushels of brussels sprouts do you think you'll harvest this year?"

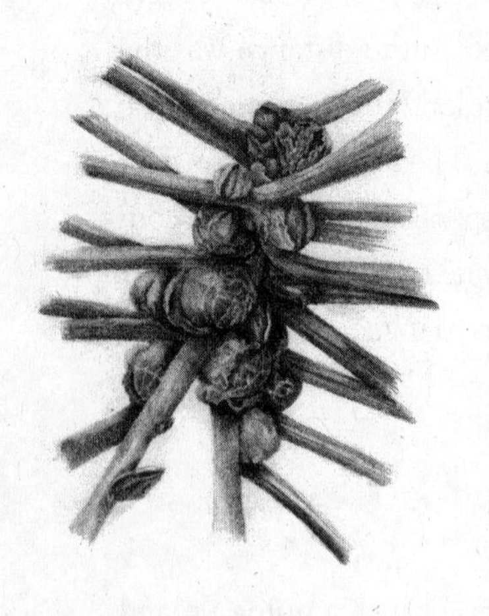

"This year so far, about nine hundred pounds, but we're not even halfway done,"* he told me without hesitation. It struck me when he responded that Daniel had a running account in his head of every jar, case, flat, peck, quart, bushel, and pound of raspberries and brussels sprouts picked and sold at Ann's Raspberry Farm.

A neighbor joined us. Daniel quickly told me his story as he was pulling in the driveway. The neighbor,

* By the end of the season the Trudels had harvested 2,100 pounds from 1,000 plants. On top of that, nearby Kenyon College put in an order for 3,000 pounds for the following year.

Dan, had been diagnosed with leukemia and hadn't worked for months. He would come by on occasion to pick berries and sprouts. He didn't pick for money or entertainment, but for health. Brussels sprouts are in the cruciferous family of plants, which are reputed to be cancer fighting, so the neighbor was trying to eat as many as he could.* According to Daniel, the last time the guy came over and picked was just before he was given a test to determine whether or not he should undergo chemotherapy treatments, but after weeks of eating brussels sprouts, his cell count was such that he didn't have to. Like all of the farmers I'd met, Daniel was au courant with the health benefits of his crops.

Daniel introduced us, and then marched us directly down to the brussels sprouts patch. While the neighbor got to work, Daniel gave me a quick lesson in how to harvest them. His pedagogy was of the tactile-kinesthetic variety, as he crouched right down and ripped into those plants, pulling off the heavy, drooping leaves, groping around the stalk with his fat fingers for the ripe sprouts, and when he found one to his liking, deftly lopping it off with his knife.

"Thees is what you want. Like thees one, see? About the same size. Big as my thumb, okay? And tight. Feel it." He peeled away a couple of loose leaves and held a perfectly round, dark green specimen between his thumb and forefinger. It looked good enough to eat. He passed it to me. "If they get too big like thees one, it's no good." He opened the fist of his other hand and

* There are a thousand articles that describe the health benefits and cancer-fighting qualities of brussels sprouts. Check this one out online: www.healthrecipes.com/Brussels-sprout-anti-cancer.htm.

held a sprout that was twice the size of the first, spongy, whiter, and full of air. "Now what I do is prepare them before I throw them in the basket so I don't have to do it again. These other guys who come and pick, they don't do it that way—they pick them, and then go back and clean them up. But you decide the way you want to do it."

At that moment a van pulled into the driveway and Daniel gave a loud shout and a wave. "That's the guys. They're picking the other patch." And off he ran.

My harvesting partner had already started picking at the far end of one wide row, so I walked down to the other end of the same row and got down to work. Before I started, I looked back at the row behind me. It had obviously been picked within the past day or two. There were scattered leaves and discarded over-ripe sprouts littering the ground from one end to the other, with big flies buzzing around the decaying debris like the aftermath of a miniature tornado. However, it was instructive, demonstrating just how indiscreet and aggressive I needed to be in order to harvest the sprouts correctly.

What an exotic plant! Huge, veiny elephant-ear leaves, with the knotty vegetable warts running up their stalk, from large at the bottom to small at the top. The soil was thick and loamy and stuck to my boots as I crouched down, leaned way forward, and felt around the stalk with my left hand. I ran my fingertips unsurely over the cabbagey lumps that progressed in size from golf-ball large at the bottom to pinball small at the top. In order to do it right, I had to get as close as I possibly could to the plant. And the longer I did it, the more intimate the experience became, so that after a while I felt like I was participating in a kind of rough and tumble animal-on-vegetable lovemaking. I had to literally fondle and grope and yank and nose and hump my groin and legs as well as my hands around those plants. Crouched low and bent forward for hours, and all the while getting slapped upside my head by those giant elephant ears, it was physically the most demanding harvest of all. Plus, it was very slow work, especially since I had decided to use Daniel's method and whittle away the excess leaves to expose the finished product before I dropped them into the basket ready for market. Not only that, but the day before I had been daintily picking raspberries, and there ain't nothing dainty and light-fingered about picking brussels sprouts. After about three hours, I had harvested a full bushel and part of another, roughly sixty pounds.

Daniel appeared at this point, slightly out of breath, to tell us it was quitting time. He surveyed my basket and nodded ever so slowly, suggesting that he was not so much impressed by the quantity but by the quality of my work. Daniel is not the type to give compliments easily. We walked up the driveway, the three of us, Daniel carrying the neighbor's bushel of sprouts and I carrying my own. Daniel, usually full of affable words, said only: "It's hard work, the brussels sprouts, hanh?" His Canuck accent was loud and clear again.

Out in the other patch, behind a row of arbor vitae, we could hear the two hired men talking loudly, laughing and singing.

"They are picking fast," Daniel added, "but they are not prepping them for market; they are just cutting and throwing them in the baskets, three, four at a time. Later they'll sit down in the shade when they are through and trim them."

We said goodbye to the neighbor and took the bushels of brussels sprouts into the garage. Allison, Ann and Daniel's twelve-year-old daughter, who looked more like sixteen, was

there getting things ready for the farmers' market the next day. With her blond hair, glasses, light blue eyes, and cheerful disposition, she was talking a blue streak, and in a very precocious manner at that, about her horse. Apparently the hoof was infected, which was causing it to limp. I could tell after listening for one minute that the kid knew absolutely everything there is to know about horses. Daniel listened and spoke with her about what she needed to do in a tone more of friend and confidant than father to daughter, and then we all went inside the house.

A quiche sat in the middle of the table as well as a tray of cheese and angled slices of French bread. The cheese was brie; it had the jalapeño raspberry jam smeared on top of it, and just standing there I could smell it. Man, what a treat! We all sat and ate together. I hadn't even been there a full twenty-four hours, yet sitting down and breaking bread was enough initiation to be considered a member of the family. The conversation at lunch started quite innocently—about Allison's horse—but soon grew in dynamics when the main point of the discussion turned to homeschooling. I sort of played the devil's advocate (with my mouth full of food), commenting about the heavy-handedness of religion in the homeschooling networks, and wondering about the long-term effects on Allison and her brother of a lack of access to sports teams, music programs, theatrical productions, and upper-level courses. Ann and Daniel fielded all the questions without a hint of defensiveness. In the end, it seemed quite clear to me that the real reason the Trudels chose to homeschool their children was that they enjoyed their time together as a family (I would learn soon another reason why keeping the kids close to home was a top priority). But after all was said and debated, proof of the homeschool pudding was bright, happy, healthy, hippologist extraordinaire Allison.

After lunch, Daniel took me to a produce auction run and operated by the Amish. It was like stepping back in time. The venue was nothing more than a large, open-walled warehouse with pallets of produce stacked in rows from one side to the other. There were cars in the parking lot out front, but the dominant vehicle was the team of workhorses hitched to large, uncovered wagons. Many of the wagons were full of pumpkins, sacks of potatoes, and bushels of corn. Men in suspenders and straw hats called back and forth to one another in Pennsylvania German. At the far end of the place they were weighing, loading, and offloading wagons; at the near end they were conducting a classic auction with a fast-talking auctioneer, chanting loudly without aid of a microphone. Under the roof were piles of produce, their prices clearly marked on top in an old-style handwriting. Daniel and I wandered around, just watching and listening, nodding hello to everyone who passed us.

Since lunch, Daniel had been rather quiet and reserved, not at all the nausea-inducing over-the-topper he'd been the day before, which led me to believe that Ann must have had a little talk with him about not overwhelming the guest. After we'd nosed around from end to end, we finally came to rest next to a tall pyramid of pumpkins, and watched and listened to the auction. The hurly-burly of it all quickly put me under a spell. The horses and the straw hats, the wagons and the old tongue began to blur the lines between the past and the present. I began to lose my sense and appreciation for the here and now . . .

Suddenly the Amish appeared before me like great sorcerers who with their magical, wide-brimmed straw hats had enchanted

the hands of time to run backward, coincidentally rendering all technological advances null and void. With them around, Bill Joy was right: "the future doesn't need us."* And why should it? The good old days look breathtakingly charming in comparison to the shifty-eyed complexities of nowadays. How pretty the horses standing there so straight and tall; how adorable the wooden wagons so simple and old-fashioned; how musical and authentic the auctioneer; how merry and bright the orange pumpkins, the purple parsnips, the yellow squashes lined up in rows . . . Yes, yes, and why shouldn't they be! Why shouldn't they indeed . . . *But how would it all work? Who would decide what new thoughts or new notions or new pleasures should be accepted or rejected? And what about the dreamers and the poets and the comedians, what would happen to them? And, after a while, wouldn't everyone look and act the same; and then, wouldn't someone start to wonder about life on the other planets, or what was so good about the good old days anyway, or how would Pamela Anderson or Eva Mendes look wearing a T-shirt instead of a plain dress and—*

Daniel nudged me, and I was roused from my neo-Luddite reverie back into the welter of the present moment. We took one last look around before heading to the van. As we were pulling out of the parking lot, I had to ask him if he would mind if I

---

* "Why the future doesn't need us" appeared in the April 2000 issue of *Wired* magazine. Bill Joy was the chief engineer at Sun Microsystems at the time. In the article, he warns that nanotechnology, genetic engineering, and artificial intelligence are the greatest threats humanity has ever faced, and he argues that we must cease and desist from pursuing research in those areas or we may face mass extinction. A *Twilight Zone*–type twist to Joy's article is Kurt Vonnegut's 1963 novel *Cat's Cradle*, which presents an eerily similar scenario to the one Bill Joy predicts, with nanotechnology turning everything into a lifeless, gray goo.

cranked up the air-conditioning, and then checked my cell phone messages, and maybe after that listened to one of his jazz CDs. He said he didn't have a problem with it, *Gott sei Dank!*

For my next assignment, I would be assisting Ann in the making of three cases of her award-winning jalapeño raspberry and chocolate raspberry jams. We drove south from Fredericktown to the "city" of Mount Vernon, where the Trudels had recently rented a commercial kitchen space for Ann to make her jams and relishes. As we drove, she told me all about the hoops the Food and Drug Administration had required her to jump through in order to get her commercial jamming license. How she had to do pit tests on all of her acidified products, that is, the relishes and mustards. In the countless hours of classes she was obliged to take, there were many tests, procedures, and equipment she had to pass, follow, and learn to use that had absolutely nothing whatsoever to do with how she made her products. She said there was an awful lot of rhetoric, too, about *food security* and *terrorism*. Oh brother!* But she's no dummy, and told me that she believes it is all just a strategy used by large corporations in collusion with the government to keep people like her out of the marketplace. She said she was the only small jam maker in the class; all the rest were employees of large, well-known canned goods companies.

The kitchen was located in a one-story brick building that

---

* Whenever I hear the word *terrorism* used by government entities these days, I tend to substitute the word *profits* for *terrorism* and find that the meaning is more accurate.

looked like a doctor's office. Inside, it was like the institutional kitchens I'd worked in during my youth. It was on the small side, about twenty-five feet by twenty feet, but it possessed all the requisite stainless-steel counters and sinks and refrigerators, pots and pans and grilled shelving units arranged solidly on top of a hard, terra-cotta-style tiled floor underneath bright, fluorescent lights. We unloaded our equipment and supplies and got right down to work.

First things first—we had to don hairnets and aprons. She started to giggle the moment she saw me with that hairnet in place, and she continued to giggle, on and off, until we were done hours later. Next she had me get the sinks ready for washing and rinsing the fancy, eight-sided jam jars. At this point, standing there in front of those three square stainless-steel sinks, filling them with hot, tepid, and cold water, respectively, it felt like I'd only gone outside to take a cigarette break from my last kitchen job. And as I washed the jars and metal tops, vivid memories of those kitchen days came flooding back.* There is no doubt in my mind that kitchen work is the magical breeding ground of all incorrigible dreamers. Dressed in checkered pants

---

* My last kitchen job was in a kosher restaurant in Milwaukee. I remember one day in particular: The restaurant manager grabbed me by the arm and said, "Hey you, come on, we've got to take these new cups and plates to the mikvah and kosherize them." The next thing I knew, I was standing up to my shins in the slush-light waters on the shores of Lake Michigan, submerging crates of new cups and dishes down in it by hand. It was below zero outside. The rabbi was there standing next to the manager; you couldn't see their faces through their frozen breath. Every few dunks, the rabbi would shake his head and point his finger, telling me to dunk again. So, my stalwart reader, if you ever want to know what kosher means, picture me, a goy, dunking plates and cups into the frigid waters of Lake Michigan.

and apron, rubber gloves and hairnets, food-soiled sneakers and food-service linen shirts, within the thrall of institutional ovens and ranges, the imagination is set free from the loserish self and soars on plumed wings to fantastical lands, far beyond the rainbow, where it frolics and swoops and loops-the-loop in dizzying dissipation for hours and hours, until the cook barks out orders for more nondairy creamer . . .

Next, she had me dicing jalapeños. And it's a very good thing that she warned me about rubbing my eyes, because I caught myself a half-dozen times just about to poke my fingertips into the aqueous humor of my itchy tear ducts. While I sliced and diced, she mixed and stirred huge pots full of fresh raspberries, sugar, and chocolate. And as we worked close together, our backs to each other, she revealed to me the story of her life. It is never a small matter when we confide in another person the contents of our joys and accomplishments, but it is an altogether extraordinary act of courage and humanity when we confide in another the contents of our haunts and fears. It would be a violation of the universal human code for me to discuss what was revealed to me in that tête-à-tête or rather dos-à-dos over a hot oven. Plus a life story never reads well squeezed inside a chapter of another person's book. But I will say that Ann and Daniel and the Trudel family are proof positive that where we go in life is not contingent upon where we came from, but rather upon where our hopes and dreams reside.

It was hard work, and when we were done I had a hefty appreciation and a whole heck of a lot of respect for what Ann did and how she did it on a regular basis. She admitted as we were leaving that she had recently hired a helper who would do the work that I did that day, but up until a few months ago, she was doing it all by herself. We managed to make two full cases of the jalapeño raspberry jam, but she ran out of chocolate for us to make a full case of the chocolate raspberry jam, so we cleaned up and packed up, and headed out. All told, about four hours' work. She let me keep the hairnet.

The next morning Daniel was up very early, and he did actually wake me up in order to get us where we needed to be on time. By seven thirty I was out in the garage, helping load the two vans. There were tents and signs and cases of jams and pickles and quarts and clamshells of raspberries and brussels sprouts, not to mention the money box. It was Saturday, and they had a full day ahead, with three farmers' markets at which to vend. I would be accompanying Daniel; Ann and Allison would leave together in the other van. But amazingly, they told me that Allison handled the smaller farmers' market in the nearby town by herself, while Ann drove another fifteen miles down the road to the third market. It was fast and furious work, getting all that stuff loaded and ready to go. These were highly motivated and incredibly driven folks. It was a wonder to behold, watching them load and pack and stack with nothing but love and tenderness and excitement for the work they were doing.

Before we left, Daniel looked at me and said, "Remember what I promised you the other day, about the raspberries?"

"Yes." I knew exactly what he meant, that all those ripe raspberries that I'd picked would have grown back.

"Go take a look," he motioned.

When I entered the hoop house I couldn't believe my eyes. Not even two days before I had cleaned the place out, totally, and now here I stood in exactly the same spot I had stood after the harvest in search of anything red—*nothing*—and two days later, that's all I saw: red dots.

"They're back!" I said to Daniel upon my return to the garage. I was expecting the same fountains of "amazing" to erupt from him again, but instead, he spoke to me in French: "*C'est incroyable*, huh?"

"O-uay," I said à la Yves Montand.

Daniel had a pretty big surprise in store for me. In conversations we had had prior to my arriving for harvest, I revealed to him that after high school I had attended college not too far from his home. Well, the farmers' market we were heading to that day was located in the very same town as my old college. And as we drove south along a meandering two-lane highway, the Ohio landscape revealing vast farms and hidden forests, wild river valleys and surprisingly high hills, I had to repress the bubbling images of long-ago faces, places, and phrases.

Around a bend, the road straightened out along with the hair on the back of my neck. Up ahead was the back entrance to my old stomping grounds. The football stadium was there, and so were the practice fields and fraternity houses up on top of the hill. In that moment I was consumed by a bright flash of understanding: this place up ahead, framed by the windshield, was a dreamscape; a familiar haunt of my subconscious mind. Certain elements of it were perhaps a bit dislocated, but overall its aura and symbolism resonated inside me like a yearbook photograph.

Daniel drove past, unaware of my startled reaction. There was a music festival going on in the center of town, so the market had been relocated a mile east to the parking lot of a church.

We were one of the first to arrive. Daniel drove to his assigned spot, hopped out of the van, and proceeded to set things up. He had his game face on. But the great thing about Daniel is his self-knowing. So as we were erecting the white tent, he said to me in a mild-mannered tone: "I know most of these vendors here, but I don't usually to talk to them. Some will come by and want to chat, but I don't encourage them because I'm here to sell my products; I'm not here to be a counselor or to talk about my children. Thees is business. Ann is very good at talking with the people, but I'm not, I'm . . . hard-core. You know?"

Yes indeed, I knew. And once we got the stuff set up, he began the hilarious process of arranging his products in a visually attractive manner. He had me fill the brussels sprout quarts so that all of the sprouts were facing one way in the box; then he had me arrange the quarts in tight rows of eight; then he had me stack them in a checkerboard pattern. On the other side of the tent, he set up the octagonal jam jars. They were displayed on little wooden tiers that, when stacked and angled, presented like toy soldiers in military formation. He looked like a man possessed, pacing back and forth, tweaking the products this way and that. At one point he completely tore down and redid two of the rows of sprouts that I had arranged.

It sort of pissed me off, so I blurted out, "Wow, Daniel, you really are French."

"Why do you say that?" he responded, his fingers diddling the quarts of brussels sprouts as if he were massaging their brainy little green heads.

"Because you obviously have their presentation problem."

His twitching fingers froze, and he looked up at me with this classic French scowl, his nose out of joint and his lips puckered:

"Presentation problem? No no! Thees is not a presentation problem. Thees is the art of seduction. You must entice the patrons with beauty and bounty. Look at them. Do they not look beautiful and worth the price I am asking them to pay?"

"They look pretty nice, but they are pretty expensive," I agreed.

"And that's how you sell—attractiveness. If my products look better than the others, I will sell more. Presentation problem, bah!" He was truly offended, as well he should have been.

For the next two hours, I watched him work his magic. He was a master salesman, saying just what needed to be said, pulling or pushing ever so gently when necessary, and yielding with highly refined grace and charm. At one point I took a stroll around. All the tents were white. All looked proud, glowing, open-for-business—both the products and the people. Farmers' markets are the purest form of capitalism. No vendor has an unfair advantage. No tent or stall is larger or more ostentatious than another. No one is allowed to advertise beyond the use of analog signs. No audiovisual effects. All the products conform to set standards. In the end, the best products are sold. I also could feel a very strong sense of camaraderie among the vendors.

After another hour or so, some of the vendors started to break down their stalls. Not Daniel. He still had half a dozen quarts of brussels sprouts and one clamshell of raspberries left. When I got back from visiting the restrooms, all had been sold, and he nearly had the van loaded.

We went into town for lunch. The whole place was set up for the music festival, with a big bandstand right smack in the middle of the street. There were beer booths along the sidewalks and folding chairs set up in theater rows below the stage. We sat outside at one of the restaurants, ate, and listened to the music. We talked about a lot of things, but mostly about Ann's Raspberry Farm, Daniel's great passion. He talked about what motivated him: "Seeing the people react and respond to our jam is the juice that keeps me going when I'm hot and tired and down on my knees weeding. To see them take a bite, and the way their eyes look at me, that's incredible." He said two other things that I won't forget: "What our farm represents is the idea that anyone can farm. If we can do it, anyone can do it." And "I don't know where it is all going, and I can't believe how far we've progressed. One thing is certain, though, we are no longer just hobbyists." It was my turn to buy, since they had been feeding me sumptuous meals since I'd arrived.

On the way back home, I asked him about the farmers' markets, and how they had evolved. Daniel said they had grown incredibly in the past ten years. He didn't have the numbers in his head, but he promised me he'd get them for me.*

As we pulled off the state road and entered Fredericktown,

---

* When I came to breakfast the next morning, Daniel had a document, handwritten, waiting for me next to the pot of coffee. Here's what it said:

- # of FMs increased 17% from 2010–2011
- FMs have doubled in the last decade
- 2011 # of FMs 7,175
- 100,000 farms sell to FMs, pumping $1.2 billion into the economy
- U.S. Gov. spends $14 billion in payments to large industrial farms, but only $100 million to local and regional farms
- 2000=2,800 FMs, 2006=3,700 FMs (22% increase), 2008=4,700 FMs (11% increase), 2009=5,300 FMs (13% increase), 2010=6,100 (13% increase), 2011=7,175 (17% increase)
- 64% of farms earn less than $10,000 gross

we passed an Amish farm crowded with buggies and horses and people sitting at long tables under the trees.

"Amish wedding," he commented.

In a field behind the barn a large group of Amish boys was playing a baseball game. They had enough players for two full teams. They had mitts and bats, and their uniforms were their own plain dress.

"Okay, Daniel, so what would they do if we suddenly stopped the van, got out, all dressed up in Cincinnati Reds baseball uniforms, and tried to get in on the game?"

He really liked that question. He sort of savored it, too, smiling as he drove, sucking on his tongue. Finally he shook his head and answered, "I don't know."

"Would they let us play?"

"Maybe."

"It wouldn't be some sort of violation, letting the English, dressed in professional baseball uniforms, play with them?"

"I don't know. But I do know that if it was a violation, someone would very pleasantly and calmly ask us to leave, and that would be it."

"Interesting . . ."

When we got back, I insisted on picking the raspberries again, and he let me. It took me more time than before, mainly because large wasps had suddenly appeared out of nowhere to invade the hoop house. They were black and white and large. Even before I started to pick, Daniel warned me that this species had a vicious sting and that I should avoid angering them. No easy deal. They were everywhere, and it was like I was dancing at arm's length

with them. When I was finished, I learned that Allison was doing research on the wasps. She would find out all about them and give us a full report.

That night dinner was exceptional. Ann's childhood friend Betsy and her husband, John Bradley, came over for dinner. John had gone fishing with his son that day, and they had caught a lot of pan-fish. He spent an hour out in the garage scaling and preparing it all, intent on serving it for dinner even though Ann had made a wonderful roast pork, glazed with her jam. In the end, he abandoned the idea, and we sat and enjoyed the gourmet meal, laughing and teasing him mercilessly about the fish. Poor guy. And I should mention that I ridiculed Daniel in front of everyone as he was preparing the condiment tray. I pointed out his French presentation problem to the rest of the people in the room as he stood there fastidiously arranging the slices of bread and cheese on the plate, meticulously spooning the relishes and mustards in neat little globs. We all had a good laugh about that, too, except, of course, Daniel. Still, he was a good sport about it. Also there was Eric, Daniel and Ann's nineteen-year-old son. He was very tall, about six foot one, with blond hair and a scratchy-looking beard. It was the first time I had laid eyes on him. A delightful young man. He didn't stay long: he had just been made manager at the local Domino's Pizza and needed to get to work.

At dinner the five adults shared two bottles of wine, and we talked for hours about everything under the sun.

The next morning I awoke to an empty house. All the action was happening outside. Ann and Daniel and Allison were hard at work in the backyard. They had set up their vendor tent back there and were preparing for a visit from the Slow Food people, who were coming all the way from Columbus to take a tour of their farm. Daniel had a huge fire roaring behind the tent to keep the bugs away. All of their jams and relishes were on display (beautifully arranged, of course), as well as heaping quarts of brussels sprouts and raspberries. Everyone was scurrying every which way. I was able to shuffle along beside Daniel as he was carrying two large pitchers of water out to the tent. I asked if I could be of help in any way, and he very graciously said, "You've been a tremendous help around here these past three days. Why don't you just relax and enjoy your last few hours here."

The people arrived all about the same time. They were expecting about twelve people. Fifteen showed up. It was an impressive group. Young people in their twenties and thirties. Highly educated. Foodies, of course. Introductions were made, and one of the group members got up and talked about Carlo Petrini and the Slow Food movement.* Then Daniel spoke, prefacing his remarks with the self-deprecating comment that he and Ann are proof positive that, since they can do it, anyone can, and then rolling right into his well-rehearsed rhetoric about their gourmet jams, relishes, and mustards "made from freshly harvested products and dedicated to the highest ideals of organic farming practices." He ending up mentioning the first prizes they had received in two categories at the 2010 and 2011 Good Food Awards, and Ann's state recognition.† Then, without warning, Daniel introduced me and asked me to say a

* Slow Food was started by Carlo Petrini in 1988.

† Ohio state representatives Margaret Ann Ruhl recognized Ann and Dan Trudel with a resolution honoring their business for receiving two 2011 Good Food Awards.

few words about what I was doing and about my book. Here's what I said:

"Don't believe this man when he tells you that if he can do it anyone can. He is a human dynamo. Everything coming off of this farm is extraordinary. And the Trudels are an American treasure."

I stuck around for a while, watching Daniel completely overwhelming folks. And when he got to the hoop house and introduced them to his raspberry plants, showing them the extra two feet of wire he had strung up to accommodate their amazing eight feet of height, he shook his head, and his eyes got big as eggs, and he exploded with "It's *amazing,* isn't it? Huh? Isn't it? *Isn't it?*"

I slipped away quietly, tears in my eyes.

Driving past the Amish houses on my way out of town, I hoped like hell to see them playing baseball . . .

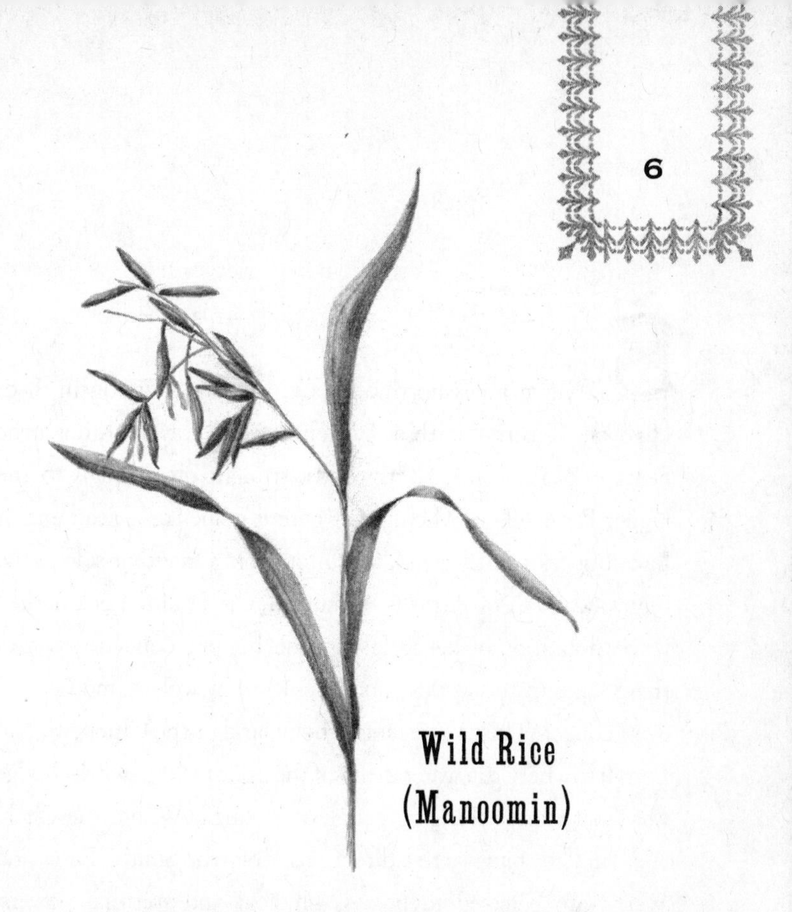

Wild Rice
(Manoomin)

LAC VIEUX DESERT BAND OF THE LAKE SUPERIOR CHIPPEWA

OLD INDIAN VILLAGE, LAC VIEUX DESERT, MICHIGAN

SEPTEMBER 2011

Sixty-five miles northwest of Green Bay, Wisconsin, U.S. Highway 45 turns north at Wittenberg and travels through the heart of Wisconsin's Northwoods, straight on through to the Upper Peninsula of Michigan, where it comes to a dead end at Lake Superior (*Gichigami*), in Ontonagon. On either side of the highway, north of Antigo, Wisconsin, the land is pockmarked with thousands of kettle lakes—the highest concentration of these lakes in the world—and populated by wolves, moose, and even cougar. This is the central homeland of the Chippewa Nation. It is where the water drum of the ancient Midewiwin Lodge was first heard calling its people to council. Where the earth, *Aki*, and its four sacred directions—North, South, East, and West—are believed to possess physical and spiritual powers. Where the creator of all things, *Gichi-Manidoo*, took the four powers and made human beings called *Anishinaabe*. And where the wild rice, *manoomin*, still grows wild.

I am connected to this land, and have been for twenty-five years through my wife, whose great-aunt Margaret (RIP), a sassy old gal with cotton-blond hair and red lipstick who smoked and drank like a sailor—all five feet of her—came to the area a decade after Al Capone and his Chicago gang had given up rum-running and hiding out in the woods. She and her second husband, Mac, a talented jack-of-all-trades, built a resort with their own four hands on the western edge of the lake they call Lac

Vieux Desert, the headwaters of the mighty Wisconsin River. She used to tell colorful stories about the giant fish they caught and the extraordinary people who used to stay at their cabins year in and year out. But woe unto him who mentioned the Indians . . .

The Indians and their fishing rights have been a bone of contention throughout the Northwoods for decades. In the 1960s, the region began to grow in popularity as people from the big cities came to fish in the lakes and vacation at the resorts.* This momentum quickly transformed the region from commercial to sport fishing. The Wisconsin Department of Natural Resources started stocking lakes with exotic species like trout and salmon, which competed against the native perch and walleye, and building dams (the higher the water, the better the fishing) without regard for centuries-old tribal fishing rights and interests. The Indians in the region opposed the tampering with water levels and wildlife, on the grounds that it violated treaties signed as far back as 1836. By the same token, sportsmen vehemently opposed the Indian practices of spearing and netting spawning fish in the spring. By the 1970s, the area was roiling with ill will on both sides. I was told that the standard fishing gear for most Chippewa was a lantern, a spear, and a gun. I once saw a sweatshirt with the logo "Spear an Indian; Save a Fish."

Today, the animosity between the two sides appears to have lost its edge, due in large part to the more passive distractions and sedentary recreations of our times, such as Cheesehead mania, large-screen televisions, *Dancing with the Stars*, the

* The area had been a favorite of President Eisenhower, who used to vacation in the nearby Sylvania Forest.

iPhone, etc. But I still remember the disapproving look on Aunt Margaret's face the first time I asked her about the Old Indian Village, just up the shoreline from her resort, and about the wild rice they were reputed to have harvested there. That frown of hers kept replaying in the back of my mind as I drove my sister-in-law's little Honda closer and closer to the reservation. Up ahead, the road inclined for a quarter mile. At the crest of the hill, the lake, dark silver with white caps, shimmered into view down below. On either side, the forest shimmied ever closer to the car as I bumped diffidently along. I could feel my heart start to pound as the road grew narrower, then turned from flat tar to vibrating, packed stone. The trees looked like they were on fire, glittering with red and gold and blush-colored leaves. Suddenly a dark shadow appeared on the hood of the car. It elongated, then compacted, showing its form—a raven, flying low over the car. It flew out in front of me, was slowly lifted higher and higher by the wind, hovered for a single moment, and then drifted behind me and out of sight . . .

"Rragh! Rraaagh! Rrraaagh! Ninizhaawandezi!" Gaagaagiw *(Raven) gave notice. ("A man from the south is coming.")*

The man was a storyteller, a traveler. He had come to gather manoomin *(wild rice) at the place where the eagles dance beside* zaaga'igan *(the lake) and beneath the* wiigob *(basswood tree). The man was alone in his smoke sled. He had come to the place to find the one they call Mitiwinaabe,**

* Roger LaBine.

the teacher and seeker of the lost mitigwakik dewe'igan *(water drum); nephew of Niigaanaash, the one who brought back the* manoomin *harvest; father of the warrior who found the eagle that danced with the people. This storyteller from the south knew nothing about the story of the dancing eagle, or the lost* mitigwakik dewe'igan *or the* manoomin *harvest, only how to drive in his smoke sled to the place. I know because I saw it.*

Boozhoo, indizhinikaaz Wiigob *(Greetings, my name is basswood tree). Now I will tell you the story of the man from the south who came to the place where the eagles dance, beside the* zaaga'igan *and beneath the* wiigob *to harvest the* manoomin.

In dagwaagin *(fall) the air and the earth are damp and everywhere the trees rattle the song of* minnewawa *(fall music), anxious for* biboon *(winter) to come so they can sleep under the soft snow. Trees have only to wait and the winter, like the water and the sun, will come and feed them and sing to them until they are fast asleep. They are the blessed ones. Life is sweet and simple. But the* Anishinaabeg *(human beings) must work hard night and day and all the year round in order to survive.*

You know that beyond the lake, the forest is full of waawaashkeshiwag *(deer). The* Anishinaabeg *must put the arrow in the bow and hunt him for food to feed his* abinoojiinhyag *(children), his* mishoomisag *(grandfathers), his* ookomisag *(grandmothers). That is not easy work because* Waawaashkeshi *has speed and agility. You also know that below the water,* Ogaa *(pickerel) has grown fat, and* Anishinaabe *must tie the bone to the string made from* wiigob *bark and dangle insects on the bone in front of* Ogaa's *nose to catch him. That is not easy work because* Ogaa *is very fussy about what he puts in his mouth. And then there is* manoomin. Anishinaabe *must paddle far out in the lake in his* jiimaan *(canoe) when the water is high and the wind is strong and the rain is bold. And you know that is not easy work. And when he reaches the* manoomin *beds he must use his push pole to propel* jiimaan *backward through the thick plants and with his knockers knock the heads of* manoomin *off the stalk and into his boat and paddle it back ashore heavy with the harvest. Certainly you know this.*

But the storyteller did not know this. Mitiwinaabe knew that the storyteller did not know this. But he needed the storyteller to help him find the lost drum. If he could teach the man from the south and prepare him for manoomin *harvest, then the man could return to his people and tell the story, and maybe they would know where to find* mitigwakik dewe'igan.

But there was no time to prepare him. Gimiwan *(rain) was coming. Not only that, but* Giiwedin *(the North) did not like this man. He did not trust him. Fortunately,* Zhaawan *(the South) liked him and trusted him. There was a great shouting match going on; back and forth they shouted across the lake at one another.* Madweyaashkaa *(sound of waves) broke loudly against the shore. Mitiwinaabe knew the man would not be able to harvest or paddle* jiimaan *in the lake with the powers so conflicted. He stalled for time.*

He showed the storyteller around the village. First he took him to the cemetery and showed him the painted spirit boxes where the souls of the Anishinaabeg *come and go between the two worlds. He pointed to Niigaanaash's spirit box. He then told him about the dam that was built for the white fishermen, which caused the* manoomin *to disappear. He told him about the restoration project started by Niigaanaash and how when the lake water was drained, per order of the Great Lakes Indian Fish & Wildlife Commission,* manoomin *steadily grew back from fifteen acres to one hundred acres after just fifteen winters.*

Next he showed him the great roundhouse, like a giant amik *(beaver) den, built of ancient white pine in anticipation of the return of the lost drum. He told him the story of the lost* mitigwakik dewe'igan, *how the western people, the Lakota* Anishinaabeg, *gifted it to the Chippewa* Anishinaabeg, *and how it had disappeared one day during the bad times. The storyteller saw the gaping hole in the center of the roundhouse, and felt a sad emptiness in his heart for the missing instrument.*

Behind the roundhouse, Mitiwinaabe showed him the ring of the four colors, symbolic of the four races of Anishinaabeg *and of the four directions of power, and where the people danced in powwow before the hunt. Mitiwinaabe recounted the story of the dancing eagle. How his son had found* Migizi *(eagle) out on the center island, unable to fly. How he had cleverly captured it and brought it back to the powwow, and how it had sat on the arms of the dancers and danced with them to the sound of the drum, round and round, imbuing them with special powers of which they continue to avail themselves.*

He explained to the man the treaties with the white Anishinaabeg, *the 1936 and the 1942 treaties and the Voigt*

Decision in 1983 regarding spearing rights, which continue to create tension between the red and white, the old way and the new way. He told about how the Sixth Circuit appellate court and the Bureau of Indian Affairs in 2002 ruled in the Chippewa's favor to lower the water level.

He taught the man about the history of the Chippewa Anishinaabeg, *and how they first came to the place; how they left the place, and later how they returned to it.*

And all the while, Giiwedin *(the North) and* Zhaawan *(the South) screamed back and forth at one another, pushing big waves up on the lake. Mitiwinaabe stalled for more time, even though there was little time left. He fed the man. He showed him the tools they would use to harvest* manoomin*—the knockers, push poles, canoes, and oars; and what they would use to prepare* manoomin *after harvest—the moccasins, the #3 wash bucket, the dancing pit, the mixing spoon, the winnow basket, the leather tarp, the birch wood. He showed him everything and he explained everything to the man.*

Then with a great shout, Zhaawan *had won the argument, and the wind grew still and the lake calm and serene.* Gimiwan *waited with* Wendaabang *(the East). Asking for guidance from and making offerings to* asemaa *(tobacco),* nibi *(water),* zaaga'igan *(lake), and* manoomin, *Mitiwinaabe took the man to the* jiimaan, *loaded him into the back, gave him a paddle, and told him to row as fast as he could straight out into the lake. It was hard paddling because the waves had not yet diminished from the fight. Mitiwinaabe and the man bobbed up and down on the surface and were not able to go very far or very fast. The waves were as high as the side of the boat, and the storyteller was much afraid. But*

Mitiwinaabe calmed the man's thoughts by having him turn the jiimaan *so that they drifted along with the waves.*

And all around the animals laughed. Some took the side of Giiwedin, *and some* Zhaawan, *and some took no side at all.*

Zhiishiib *(Duck) said, "Look at the storyteller, ha ha ha. He is afraid. And he should be with his pockets full of wood-made things, smoke sled jewels, and leather pouches. He is out of balance. When* Giiwedin *starts to blow again, he will surely fall over the side, and with his pockets so full sink to the bottom of the lake."*

Maang *(Loon) said, "No, you are wrong. He paddles well. Besides, he has been here before. I have seen him around, poling for* Ogaa *with his long hair and long arms and big hands."*

Wazhashk *(Muskrat) said, "You are both wrong. Do you not know who this is? This is* Iagoo, *the Big Boaster. He is an absurd one. He is only pretending to be afraid, but he is up to something, I can assure you."*

Amik *(Beaver) said, "This is not* Iagoo. Iagoo *is old and black and twisted like the bark on* giizhik *(cedar). His eyes are bigger than his ears, and his ears are twice as large as any other man's. And his legs are like the trunks of beech trees and his arms are as strong and as dense as* maana-noons *(ironwood). Besides,* Iagoo *only comes back when the moon is blue, full of incredible tales, like the time he visited a lake where the mosquitoes had wings large enough to use as sails. And where serpents with long manes like horses swim on top—"*

Maang *(Loon) said, "*Amik, *you fool, you have been listening to too many stories told by the crows."*

And as the animals talked and debated, the two Anishinaabeg entered the warm, shallow zone of the manoomin bed. Giizis (sun) appeared briefly to smile at them in approval. Immediately Mitiwinaabe stood in the front of the boat with his feet and legs pressed fast against the sides of the jiimaan. And with his push pole, he turned them slowly around, and commenced to push them backward through the tall, ripe stalks. He gave orders for the storyteller to work the knockers by pulling the heads down with one knocker and hitting them with the other knocker, distributing the dried seeds into the bottom of the craft.

At first it did not go well. The storyteller could not get it right. But then, suddenly, he felt it, and as if he had done it a thousand times before, he was knocking the manoomin into the boat, left side, right side, left side, right side, like the sound of the rain playing softly on the bark.

For hours they gathered manoomin, Mitiwinaabe pushing and the storyteller knocking. And when Mitiwinaabe had taken them all the way through the hundred acres, he again turned the jiimaan around and together they paddled hard, back toward the place beneath Wiigob. Gimiwan had arrived by this time, and the argument started by Giiwedin and Zhaawan and taken up by the animals of the lake had provoked the two original antagonists back into debate. And so the waves on the lake began to swell, but just before the two got to shouting at one another, Gaagaagiw's ominous shadow appeared above, and he gave notice:

"Raagh! Raaagh! Zhaagnaash Bawshikinaway jiimaan, majimashkiki! Inashke!" (Pale faces in big boats with ill intentions are headed straight for the two in the canoe. Look!)

All of the animals stopped their debate and turned to look. Giiwedin *and* Zhaawan, *too. It was an evil scene. The two* Anishinaabeg *in the little jiimaan, paddling on the big wide lake, and the big, broad smoke boat full of white* Anishinaabeg, *the bow so tall that* Mitiwinaabe *in the front could not see the faces of the men coming toward him. He stood and waved his arms, but the big smoke boat kept coming straight and fast. The storyteller in the back looked up and gave a great shout and tried to steer hard away. They would surely be hit!* Amik *hid his eyes.* Wazhashk *dove down below the water with* Zhiishiib *next to him.* Maang *lifted up in the water with his long wings and barked in protest, "Kwuk Kwuk Kwuk!"*

At the last possible moment, the big smoke boat turned astern, and the pale faces came into view, full of smiles and laughter and jeers. But what followed them was equally as treacherous as the big smoke boat itself: a great serpent's tale of wake water, rolling over itself and growing fatter, slashed and rolled toward the two in the tiny jiimaan, higher and higher. Mitiwinaabe in the front, with the oar, pulled the jiimaan *straight into the wall of water, just as they were about to be hit on the side and capsized. The tiny jiimaan lifted higher and higher in the water, gave a hop, then plunged down fast and hard. The storyteller in the back dropped his paddle and held on to the sides with both his hands. Three times the little jiimaan was lifted high, and three times it fell back into the water hard and true. But the two remained safe and secure inside.*

When the danger was past, Mitiwinaabe gave a motion with his hands to indicate agoozo *(balance), and without*

exchanging words or glances, they paddled again toward the place beneath Wiigob, *with* Giiwedin *gently blowing them from behind and* Giizis *softly smiling above.*

That is how the storyteller came to know about the manoomin *harvest at the place where the eagles dance, beside* zaaga'igan, *beneath the* wiigob. *As for the return of the lost* mitigwakik dewe'igan? *Mitiwinaabe waits.*

It was a little after 9 a.m. the next day. I was accompanied by my wife, her parents, and her sister. We arrived in two cars. Roger had prepared a roaring fire and laid out the wild rice we had harvested the day before on a big blue tarp and rice from a previous harvest on another. The problem was the weather. It was drizzling. He had a little event tent set up, under which the fire burned innocently. Next to the fire were all the tools of the trade, namely a big #3 wash bucket, three sets of push poles, wooden knockers, mukluks, a leather blanket, a birch bark tray, and an old oar.

I introduced everyone, and Roger, being the consummate teacher, got right down to business by having us gather around the array of implements. It was a bit of a letdown for me to begin the day like this, lecture style, because the day before had been such an exciting, albeit dangerous, experience with just Roger and me out on the lake, trying to survive the waves, weather, and vindictive fishermen. Not only that, but here we were among the entire Cheesehead nation. My in-laws—my wife, and sister-in-law, too—are classic burghers from Milwaukee and huge Green Bay Packer fans. Of German-Irish lineage, the parents-in-law are both former teachers; my sister-in-law is a hospital administrator, and all, including my wife, comport themselves as emissaries of a gentle, dignified, amiable, thoughtful, and honest-to-a-fault nation of *Volk* (except when in the presence of volks from Minnesota). And I am delighted to say that even with their great hoary heads of hair, dressed in Packer colors, the parents-in-law were not out of context standing there in the Old Indian Village, at the place where the eagle dances, ready to help process the *manoomin*.

Before describing the equipment, Roger gave us a little background and natural history lesson. He first told us about his uncle, called Niigaanaash, who with support from the Great Lakes Indian Fish & Wildlife Commission first began a wild rice restoration project back in 1989. Since that time, the rice harvests on Lac Vieux Desert have grown. In 2006 Roger hosted his first rice camp, with 105 people in attendance, half of whom were tribal members.

As for the plant itself: *manoomin*, which means good berry in Ojibwe, is an annual plant that germinates under the ice. Right after ice-out, shoots sprout up to the surface of the water. When it breaches the surface, it goes from being an anaerobic to an aerobic plant, and of course at that stage it is quite vulnerable to any and all agitation. It's like a floating leaf, he described it, and rising water puts tension on the root system—too much and it snaps. Spring runoff is a crucial time, and high water can easily kill the plant. The dam, of course, is a huge problem, but since the restoration project began in 1989, the water levels have been kept low and the rice has come back (to the chagrin of many a sportfisherman from downstate).

Next, he explained each piece of equipment's function and what it was made out of. Starting with the long push poles: he told us about why you must push rather than row through the

rice beds (the plants are too thick and the rowing would injure or destroy the rice). The push poles are twelve feet long and have forked ends made from cedar, one of the four medicinal plants of the Chippewa. The forked ends, which look like they could double as heavy-duty slingshots, are tied in place with leather twine; their dual ends keep the roots of the rice from wrapping around the pole. Next, the knockers are also made of cedar, and look and feel exactly like giant drumsticks, but are not so shiny or perfectly lathed. They are used in tandem—one to pull the stalks over the edge of the canoe and the other to knock the rice into the boat. The rest of the equipment—the moccasins, winnow basket, and #3 wash bucket, he explained—would be self-explanatory once we got to using them. I didn't ask about the #3, but assumed that if one goes to the hardware store and asks to buy a wash bucket, the response from the sales clerk will be, *Which number?*

Curing the rice. — In more traditional times, the rice was dried on birch bark or blankets spread on the ground and continuously aerated and sun-dried for two days. About the only nontraditional element to the entire rice harvest was the two plastic tarps they were using to dry the rice on.

Parching. — Ever since the arrival of the white man, the cast-iron kettle and later the wash bucket was used to parch the rice instead of the two-day drying method. Like all new inventions, it's faster and easier, with more consistent and predictable outcomes. He then demonstrated the process by filling the bottom of the bucket with the rice, about an inch thick. He then placed the bucket at an angle over the fire and began swishing the hulled rice around and around with an old canoe paddle.

Laborious work. He said the rice is properly dried when the kernels snap in half between your fingers. Old hands can simply tell when it's ready by the rich, nutty aroma. He had my father-in-law do the swishing, then my mother-in-law, and then the rest of us. It took about twenty minutes for that aroma to get just the way he wanted it.

Jigging or dancing on the rice. — After parching, the rice is removed from the hull by dancing on it in soft moccasins. Behind us, outside the tent, was a circular hole dug in the ground, about a foot and a half in diameter, with a leather cloth covering its uneven contours. For balance and to help the dancer control the amount of weight applied to the rice, Roger had two poles placed alongside the pit in a V-shape, like stair rails, set up on either side to hold on to. This was the rice-dancing platform, and with his white moccasins on, Roger demonstrated the dancing technique. As a teacher, Roger likes using analogies. Before he began to dance, he asked us who invented the Twist, and we all shouted "Chubby Checker!" Sure enough, that's what he did, swiveling his hips and shaking his butt, twisting and turning back and forth. We all got to laughing. He didn't mind. Chubby Checker is a funny name.

Winnowing. — The final step is winnowing, or tossing the rice in the air to separate the kernels from the chaff. Roger placed two handfuls of the dehulled rice into a *nooshkaachinaagan*, or birch winnowing tray, and with a downward shaking movement, the lighter chaff would blow away, leaving the rice kernels ready for cooking or long-term storage. It took about three minutes of constant shaking per basketful.

Tweezering. — And as if on cue, Roger's brother Paul arrived atop his noisy and smoky ATV. I'll give you one guess, my gifted reader, who Roger's brother Paul looked like. You got it—Chubby Checker. Amazing! Roger introduced Paul to everyone, and with all seriousness told us that Paul was the Tweezer Meister. And so Paul, armed with tweezers and magnifying glass and thick reading glasses, sat down and proceeded to meticulously remove tiny bits of chaff from the protein-rice grain, making it ready for sale in the marketplace.

All told, from parching to tweezering, it took about an hour per pound of product, which doesn't include the time and danger of harvest.

To participate in a harvest ritual like this, which dates back a thousand years or more, is to experience firsthand the purest harmony with nature. Wild rice requires no maintenance, no machinery, no artificial fertilizers. You do not even need to plant the product; it plants itself and is ready to be harvested, pure and natural, year in and year out. In California, the world's largest producer of wild rice with some twenty thousand acres and fifteen million pounds produced each year, the harvest process is completely mechanized. As a result, the cost is half what you'd pay in the traditional growing regions of the upper Great Lakes. But here's where the consumer becomes the cultural steward and environmental champion. To buy this product at its natural price, where it was harvested, supports in every way indigenous culture and families; it supports sustainable agriculture and sportfishing; it supports the Great Lakes region and its communities; and it supports *ninoododadiwin* (harmony), *agoozo* (balance), and *manajiwin* (respect). Buying *non-wild* wild rice is a lame, limp-wristed slap in the face to traditional *manoomin* harvests and harvesters.

Once everyone had had a turn trying his or her hand at processing the *manoomin*, it was time for them to go. The in-laws were scheduling their return to Milwaukee in time to see the Packers play the Carolina Panthers. My sister-in-law was driving to the western part of the state to visit her son at college. Even my dependable, musical wife was committed to returning to the resort home to help her mother tidy things up. So there I was again with Roger, looking at a pile of work to do under an increasingly dubious sky. There was a whole tarp and a half of *manoomin* to process, and I had told him that I was prepared to do whatever work needed to be done, so that's just what I did for the next four hours. His uncle Charlie had joined us, and he was pretty much in charge of the parching process.

I was a dancing fool! With my big feet stretching out those lovely white moccasins, I twisted and shimmied and swiveled until the balls of my feet were raw and swollen. And it seemed that every time I looked up into the sky, a huge raven was there soaring low over the treetops. There were bald eagles, too, high above the lake, a mile in the air, watching all. So under their watchful eyes, I dehulled and winnowed every ounce of *manoomin* that Roger had laid out.

When we were done, Roger drove me in his truck up to the sweat lodge, which stood behind his uncle's house. It was a round, igloo-like building, as primitive as it gets—erected of bent tree limbs and poles and covered with blankets, tarps, and animal skins. Inside it was dark and musty, with a hard-packed dirt floor, and a foot-deep hole about three feet in diameter cut into the middle of the space. This pit is where heated rocks

called *mishoomisag* (grandfathers) were placed and provided the heat for the sweat. As we stood next to the pit, Roger proudly explained its purpose and religious significance. They come to the sweat lodge to work out their problems. They sit in total darkness around the hot rocks and consult their grandfathers. It was a very powerful moment standing there in that earth house. In college I had lived and studied for a semester in Florence, Italy. I took all the tours of all the magnificent churches: San Lorenzo, Santa Maria Novella, Santa Croce, Il Duomo. Masterpieces all, but in my estimation they don't hold a candle to the sublime religiosity of that holy place in northern Wisconsin.

Before I left, Roger gave me not one, not two, but three full bags of *manoomin*—more than double what he and I had harvested the day before. I had a lump in my throat . . .

And that night, back home in South Milwaukee, my mother-in-law fixed one of her classic midwestern dishes: round steak and rice. And yup, you guessed it, the rice was none other than *Minute Rice*.

Cranberries

BRENDA COBB CRANBERRIES

MIDDLEBORO, MASSACHUSETTS

OCTOBER 2011

It's amazing how dramatically the northeastern landscape changes as you motor across it. Even tiny little Massachusetts explodes into gorgeous, rocky-topped mountains as you climb up from the Hudson River valley at its western border with New York. Then, rather quickly and dramatically, the mountains and cliffs full of steadfast oaks and resplendent maples and muscular beeches give way to loam-filled plains full of birch, hemlock, and white pine. And then, even more quickly than before, that all recedes and is replaced by sandy soil, full of pitch pines and scrub oaks and sugar maples and cranberry bogs.

Southeast of Boston, especially this side of the Cape Cod Canal, cranberry bogs are everywhere. Around just about every bend, you come upon a bog, sometimes two, sometimes a dozen in a row. On either side of the road, these flat, mossy-looking, rectangular fields are dotted with colored flags, edged with deep trenches, and guarded by sentry boxes* at one end or the other. In the fall of the year, many of these bogs are intentionally flooded. Once they are underwater, strange machines roll over the top of them, setting loose the ripe-red berries, which float to the surface and express themselves in a dazzling artistic exhibition that I would call "Variations of Roses and Reds." Cranberries are one of only three commercial fruits native to the United

---

* Actually pump houses; they just look like sentry boxes.

States and Canada, along with blueberries and Concord grapes, and they provide not only a unique culinary experience but also one of the best sources of antioxidants on the planet, not to mention a pretty dependable source of income for a small, job-challenged region.

According to my wife, I had typecast her as a berry picker. Not that there's anything wrong with that. She was with me on a few of my excursions to pick blueberries; she also helped me dance on the wild rice, the Chippewa's *manoomin* (good berry); and now here she was accompanying me to old Cape Cod, ready, willing, and able to harvest cranberries. It seemed a logical match. Cape Cod, after all, is famous as a romantic getaway spot. There definitely is something about the salty air and the lapping waves upon the shore, and the pointed peninsula where the Pilgrims landed, if you catch my drift. The literature will back me up on this.*

It was rainy and windy when we arrived at Brenda Cobb Cranberries, a small, family-run, five-acre bog. Brenda Cobb, the owner, reminded me a lot of Doris Day (especially in the movie *Send Me No Flowers*), and the first thing she did was take us to her garage and show us her two cranberry harvesters. They looked exactly like giant sideburn trimmers on wheels, permanently set at 45 degrees and with handles. She prefaced her show-and-tell by telling us that all cranberrymen are mechanics and inventors. It's a simple matter of economics, she explained, because the industry is too small for large industries to bother

* If you want to read something over-the-moon romantic about Cape Cod, check out the italicized section of Richard Russo's *Empire Falls*, part 1, chapter 8.

manufacturing expensive equipment. Almost all of the machines used in the cranberry harvest are custom-built by cranberrymen. Her own father had made the two harvesters she owned. And an hour later, when her dad, Larry Cowan, appeared, he pointed out the Kohler engines to me, with the comment that nowadays cranberrymen make their harvesters with a four-horsepower Honda engine:

"They start with one or two pulls, not like these Kohlers. Heck, you can pull all day sometimes and they won't start. But once you get them going, the Kohlers are workhorses," he said, talking to me as if I knew something about engines and brand names and horsepower, which I don't.

Brenda's son Tristan and his roommate from college, Ben, were there with their shirts off (it was a balmy day) and so full of youth they were practically floating. They had come home for the weekend to help harvest. However, it was too wet to "dry-harvest," Brenda told us, frowning at her bog behind us. I felt a pang of dread creep in, thinking that I would be missing out on another harvest, since these were the only two days I could spare. But no sooner had she said that and frowned than she brightened and offered something else:

"But my ex-husband, Gene, has a bog five miles down the road, and he's wet-harvesting today, so we can go over there and help him out if you'd like."

"Sure," I effused.

Without further ado, we hopped into our vehicle and followed Brenda in hers through the cranberry countryside, over to the Burgess Bog Company. We entered the property down an overgrown dirt road in the middle of a suburban neighborhood. It was very odd. We literally drove between two Cape Cod–style houses with toys and bikes and swing sets in the yard, through a thin stand of tall pines, and came out onto a vast bog. Brenda slowed down her little Honda and continued toward the back of the acreage, where a truck and some tall water machinery were visible off in the distance.

We drove slowly along the raised trails between innumerable bogs, under a gray and drizzly sky. It was an otherworldly feeling, bumping along a narrow dirt path between two flooded bogs full of red cranberries; it felt like riding on the painted frosting of a cake. Eventually we came upon a man seated on a strange motorized contraption, almost like a go-kart welded onto an extra-wide frame with big round tires and pulling along behind it a dozen spinning saucers set horizontally half in and half out of the water. By golly the whole scene was like something out of a cartoon, and it took my wife's total recall ability of childhood books read, cartoons watched, and jazz standards and Broadway musicals sung to nail it on the head: "He's a Who down in Whoville."

The man on the loupadoodler, as we called it, was Gene Cobb. And what his machine, called a reeler, was actually doing was "reeling off" the berries. Gene Cobb is an amazing physical specimen for a man of fifty-three. At six foot five, with strawberry-blond hair, sky-blue eyes, washboard abs (he went shirtless the entire time), and giraffe-long neck, he has been a cranberry man for thirty years. Gene owns something like fifty acres of cranberry bog. He is a full-time cranberry man and strictly a wet harvester. Wet-harvested berries are only sold for juice and cranberry sauce.

Gene was a perfect example of the cranberry man/inventor. All of the wet-harvesting mechanisms that he was using he had built himself. The reeler, for example, is a four-wheeled contraption that rides softly over the berries once the bog has been flooded with water. From the shore, it looks like he is riding on top of the water, but in fact he is driving across the flooded bog on top of the cranberry bushes. With the sifter or reeler in the back churning as he drives, the ripe berries are gently dislodged from the bush. The trick is marking your spot with flags in the flooded bog water so you know where your line is and what you've already reeled off.

After that, the process gets even more interesting. The berries, once reeled off, float to the surface. Cranberries are designed to float—they have four air pockets under their airtight skin. The wind usually pushes them to one side of the bog or the other. If you are an art enthusiast, you might be inclined, like I was, to stand back and look for the subject to appear in the pond-sized, rose-red, Seurat-like pointillism before you. It's a dazzling impressionistic effect with infinite variations of reds and yellows delighting the eyes. I had a fleeting thought that maybe Seurat had visited the Cape and come up with his idea for pointillism after experiencing a wet cranberry harvest. Anyway . . . a floating boom (an item that is now manufactured and not custom-made) was set around two sides of the bog, and once all the berries were afloat, it was slowly tightened around the crop and pulled tight. Gene's assistants on this day were his two little neighbor kids, Mikey and Harry. Both were about nine years old. They were running around like wild Indians, and I was impressed with Gene that he had enough confidence in them that they wouldn't somehow injure themselves. Harry was on the small Kubota tractor that had the end of the boom attached to it. It was his job to back the tractor up, which then tightened the circle of the boom around the berries. Again, to experience a huge circle of red cranberries edged in black moving around a pond is an extraordinary thing. Wild nature, pulled and corralled and cinched by the energies of humankind, can really put asses in the seats. My wife thought it looked like a huge pink comment balloon.

The other great device used in the wet harvest was a spectacular two-story affair. I thought it resembled a Rube Goldberg machine, but, as usual, my wife had the much better analogy, as she was convinced that it was one of the same contraptions featured in *How the Grinch Stole Christmas*, specifically, the Great Who cardio Flux Machine that the Whos down in Whoville like to ride and play during the holidays. The Great Gene Cranberry Sux Machine, as we came to call it, had a second level and more length of hoses than a fire engine. There were hoses for input and hoses for output, and one big schnozzle that sucked the berries up, sending them into the machine and out to various destinations. The most amazing part of the apparatus was beyond view. Gene explained that there was a slatted component inside the thingamabobber that separates the good berries from the bad ones, the deflated from the buoyant—the good going into the bed of one truck and the bad into the bed of another.

At one point, standing up top and looking out over the acreage, Gene sidled over to me and pointed out the numerous pump houses stationed around the bog. This was how he flooded the place. With a series of pumps and pipes, the water from one bog, once the reeling off was complete, could be drained and then pumped into the next bog, and so on down the line. He had recently had new pipes put down that would last a hundred years.

I got the sense that Gene was able to, and probably often did, harvest his fifty acres of cranberries completely by himself. But, like with all farming, the more the merrier. Usually one or two people with rakes (eight-by-eighteen-inch rectangular plywood blades held perpendicularly at the end of a handle) could do the job of pushing the berries into the suction hopper. One man per truck would also deliver the product to market.

Mikey and Harry were hysterical. They were as much nuisance and hindrance as they were help. They just wanted to ride and climb around on all the cool machinery. I was blown away by Gene's patience, as those kids were feisty and hyperactive and constantly up and down the ladder of the Great Gene Cranberry Sux Machine with all those moving parts . . . Eventually Tristan and Ben arrived and took over for Harry and Mikey. My wife and I did offer our services when it was apparent that Harry and Mikey's interest and productivity were flagging, but the two young men, wide-eyed and shirtless, just seemed too perfect for the job. They donned waders, wielded rakes, and got right down into the bog to rake the cranberries toward the sucker.

Meanwhile, we busied ourselves with capturing it all on digital photography from every possible angle. At one point, as I stood high up on the very top of the Great Gene Cranberry Sux Machine, watching the berries entering the intake section down below and then watching them, in defiance of gravity, spew from two different effluent pipes ten feet above where they'd entered, I decided that the manifest genius of humankind in the form of machines, like "the predictive power of mathematical physics," is one of the most awesome things in the universe, and that we should be constantly in awe of both, "not merely of the universe, but also of the men and women who have stripped, and continue to strip, that universe of its mysteries—and do so without diminishing the wonder of it all."*

When things seemed to be well in hand, and everyone was literally up to his waist in cranberries, Brenda took my wife and me aside and asked if we would like to go into town and get a late lunch. We were hungry, so we agreed. Bidding farewell to Gene and all the other Whos in Whoville, we once again followed along behind Brenda. This time we drove into the town of Carver, known as the Cranberry Capital of the World. Brenda brought us to the farmers' market, which was effervescing with New Englanders, and led us directly to her mother Louise's booth. This was a nice surprise. Louise was selling goat cheese, and we bought a heaping helping of it. After our purchase we stood and chatted with Louise, who assured us that she would join in the dry harvest the next day, so long as the weather cooperated.

We bid her adieu and wandered around. It was an odd farmers' market because it was under a three-acre stand of old-growth pines that soared 120 feet in the air. The ground was dry-packed sand and it felt at times like you were walking around on a giant's scalp, under his soaring hair follicles. Music was playing. Lots of good smells wafted through the area. Kids ran about unsupervised. The funny things was, we couldn't find a single cranberry for sale!

After visiting nearly every vendor in the place, we decided to sit and eat pizza at the pizzeria across the street. There the three

---

* From "Higgs ahoy! The elusive boson has probably been found. That is a triumph for the predictive power of physics," *Economist*, December 17, 2011.

of us sat, in a window booth, getting acquainted, when what to my wondering eyes appeared? Gene! And he had a shirt on. He came in cowboy style—bowlegs and blue jeans, thumbs dipped in his belt—with his new girlfriend. They had called in an order for pizza and were taking it back to the bog to eat. He stood there towering over us, bent at the hip, leaning on the table, talking in his radio-announcer's voice. He had a waggish wit and punctuated his punch lines with a dramatic toss of his head and a *yuk-yuk* of a laugh. I could tell my wife was very much taken by the whole scene. I noticed that the two women, Brenda and Gene's girlfriend, seemed to know each other, and there wasn't the least whiff of discomfort or embarrassment between them.

When they left, Brenda filled us in on Gene's background and the Burgess Bog, which had been in his family since the 1890s. She described a 1940s scene: a man in a starched and pressed white shirt and tie, with his sleeves and pants rolled up, raking cranberries in the bog after a long day's work. That was Gene's great-uncle. Another interesting tidbit we learned was that Gene's brother is a cellist for one of the Lincoln Center orchestras, and Gene's mother is a master violinist who is well-known throughout the area for her private instruction. Gene himself was quite an accomplished cellist, too, but had decided nearly thirty years ago to become a cranberry man instead. Brenda did not talk about the dissolution of their marriage, only saying that they continued to have an amicable relationship.

The next morning, we arrived back at Brenda Cobb's bog ready to dry-harvest cranberries. And we did just that, commencing right at 10 a.m. and concluding at 4 p.m. It was a full day's work, and Brenda, from the glowing statements she made regarding our workmanship, was quite glad to have us there. It was a clear, crisp fall day. No rain in sight, although there were plenty of fat cumulus clouds marching overhead, but they were the friendly kind, the very same that have inspired artists through the years.

Dry harvest is a heck of a lot less mechanized than wet harvest, demanding truckloads' more elbow grease and manual labor than mechanical engineering. The equipment? Those two push-behind harvesters, scores of burlap sacks, a wheelbarrow, and a tractor with a little flatbed trailer. They are carried to market in three-foot-square-and-tall bins. There is a strainer that is placed on top to prevent the cloying vines from entering the storage bins. The harvesters, as I said, look and work like sideburn trimmers. The teeth trim, side to side, through the berry plants, pulling the berries off the vine and upward via a conveyor into a burlap sack hooked in place under the handlebars. The person wielding the harvester is obliged to trim with the vine, not against it. Brenda had me work the harvester almost from the start, and it didn't take me too long to get the hang of it. The trick was finding and following in the wake of the harvester in front of you, and only trimming half a row—about eighteen inches per pass. If you tried to trim off more than that you would miss too many berries. Plus, the teeth got clogged easily with vines. It was slow, arduous work. On top of that, the burlap sacks filled up very quickly. I kept forgetting to change the sack under my hands, and by the time I remembered, my sacks were bulging with berries—seventy or eighty pounds each, when they should have been no more than thirty or forty. I learned the error of my ways when Brenda took me off the harvester detail and put me on sack collecting.

Larry Cowan, Brenda's dad, showed up before lunch and hopped right on the tractor. Larry first bought the bog in 1978. When he retired in 2002, he turned it over to Brenda. With his gray hair, light blue eyes, and rimless glasses, Larry was in good shape for a man close to eighty. Medium height, but on the slim side, he was a classic New Englander who spoke little, but when he did it meant something.

Whenever we finished a round, and before heading back out to the bog, he would recount a little anecdote about the old days. He told me about the Cape Verdeans who were the original pickers back in the day, before there were any harvesting machines. He went inside and came back out with a traditional cranberry picker. It was quite an ornate contraption. From a distance it looked like a wooden basket, big as a bread box, but close up it had long tines (five-sixteenths of an inch apart, I was told), thick and durable, dozens in fact, and a handle on top and on the end. The cranberry pickers would sweep it through the vines in a sideways motion, sort of like throwing a pail of water, then empty the contents in boxes as they went. He told me the story of one old-timer who could pick a hundred boxes a day with a traditional cranberry basket.

Every so often, great blue herons drifted by, over our heads, checking out the bog for frogs and little animals that the machines and people might have disturbed. Larry and I worked together most of the afternoon, picking up the burlap sacks left on the bog by Brenda and her boyfriend, who had showed up to help harvest. Larry drove the tractor and I followed along behind, heaving those sacks of berries onto the flatbed. I felt a lot like Lennie in *Of Mice and Men*, lumbering along, picking up bags, and tossing them onto a flatbed. Backbreaking work.

With about sixteen sacks maximum piled on the back of the tractor, Larry would drive the load back over to the bins that sat on a long, semi-trailer-sized flatbed. There the ladies waited for us: my wife and Brenda's mom, Louise (the goat lady), and Larry's ex (she showed up, too, shortly after lunch). It was interesting to watch and to listen to the two exes interact. Occasionally Larry would say something sarcastic, and she would fire back something equally biting. It was all just good fun. I helped the ladies lift the bags and dump them, one at a time, onto the filter screen that fit over the tops of a bin. The screen did a good job of keeping the unwanted green vinery out. This unwanted vinery was then tossed onto the ground, to be composted somewhere on site later on, into an ever-growing pile. Louise was a medical marvel. Who knows how old the lady was. It didn't matter: she outworked my wife and me, lifting burlap sacks of berries, working the berries over the screen, and then racing out into the bog and reeling off the edges of the ditches by hand with a traditional basket (you can't get the berries on the edges with dry harvesting the way you can with wet harvesting). As I said, wet harvest is done strictly for cranberries bound for juice or sauce, but dry-harvested berries are strictly sold as fresh fruit by the pound in clear poly bags. About 10 percent of Massachusetts cranberries are dry-harvested.

But at one point, both Louise and Larry really pissed me off. We were talking about the unseasonably warm and wet weather when Larry up and asked me, "You know who created global warming, don't you?" I had no idea what he meant by that. "Al Gore," he quipped. And immediately Louise added, "Yeah, all this nonsense about global warming. It's just weather patterns: you have warm years and not so warm years; dry years and wet

years. Global warming . . . bagh, what nonsense! If you ask me, it's all just politics."

I was going to say something pretty nasty, like: *Easy for you old farts to say because you get to skip off the stage blissfully ignorant and arrogantly proud that the planet is just hunky-dory. But it's not. After you're gone, my generation and my kids' generation will be stuck cleaning up all the nuclear waste and greenhouse gases and heavy metal contaminants and chemical fertilizers and plastic bags, etc., that your generation created.** But then I thought about Wes Jackson and Wendell Berry, members of Larry and Louise's generation, those intrepid generals in the ongoing war to save the environment. Out of respect for them, I held my tongue.†

By day's end we had filled sixteen bins, or about sixty barrels, which, we were told, is roughly 6,600 pounds of berries. It was hard work, but it was robust work, and satisfying. Heck, Brenda

---

* At the time of the writing of this very page, a recent discovery was made in the Arctic Ocean by Russian scientists: "Dramatic and unprecedented plumes of methane—a greenhouse gas 20 times more potent than carbon dioxide—have been seen bubbling to the surface of the Arctic Ocean by scientists undertaking an extensive survey of the region." http://www.independent.co.uk/environment/climate-change/shock-as-retreat-of-arctic-sea-ice-releases-deadly-greenhouse-gas-6276134.html.

† In the coda of his latest book, *Consulting the Genius of the Place*, Wes Jackson writes: "There is the human population, but there is also the population of deep freezers, the population of houses, the population of pop-up toasters, the population of automobiles. All members of these populations occupy space, and all are dissipative structures, which is to say, all are subject to the second law of thermodynamics, the entropy law. We need to reduce the numbers of many of these populations, including humans. That's number one." And right alongside that, a story about the leaders of the island nation of Kiribati, who are considering moving the entire populace to Fiji, fearing that climate change could wipe out their entire Pacific archipelago.

told us she'd get a dollar a pound for her product, and she still had another fifteen thousand pounds to harvest. Do the math. That was her gross for the year. And as we were leaving, Brenda gifted us a heaping bag of cranberries. It felt like Thanksgiving and Christmas at the same time.

So, with the cranberries reeled off and trimmed back and packed into crates and loaded onto flatbeds, my wife and I took leave of Brenda Cobb Cranberries, singing a tuneful reprise of "Welcome, Christmas" as we bumped along out of town:

Fah who for-aze!
Dah who dor-aze!
Welcome Christmas,
Come this way!
Fah who for-aze!
Dah who dor-aze!
Welcome Christmas,
Christmas Day . . .

Potatoes

WOOD PRAIRIE FARM

BRIDGEWATER, MAINE

OCTOBER 2011

A few weeks after graduating from college, I went up to Maine to hike in the woods like Henry David Thoreau once did. With a childhood friend as my hiking companion, we started our adventure at the New Hampshire–Maine border, bent on hiking all the way to Mount Katahdin, the legendary mile-high terminus/starting point of the entire 2,170-mile Appalachian Trail. It was a mere 280 miles north and east over the tallest mountains, through the densest forests and across the wildest rivers imaginable. *Hoorah for miles!* After losing the trail a number of times, nearly being struck by lightning on a treeless mountaintop, and getting swarmed by black flies and mosquitoes almost constantly, we managed to advance 175 miles before one of my boots gave out. We could not continue and had to turn back. To this day, the Maine woods is the wildest place I have ever experienced, and it remains an affirmation of Thoreau's belief that "[t]he most alive is the wildest. Not yet subdued to man, its presence refreshes him."*

And it was with that expectation of wild aliveness that I sat looking at the wilderness outside the car. My traveling companion this time around was my oldest daughter, Kather, the original "loser" who, with her parents' encouragement, had recently dropped out of college to pursue her dream—becoming a stand-

* From Thoreau's "Walking" (1862).

up comedian.* It was late morning in early October as we drove north along Interstate 95, somewhere north of Bangor, Maine. Out there, beyond the trees, some forty miles to the west, was my white whale, Mount Katahdin. It was raining and misty and gray and I couldn't see it. We were the only vehicle on either side of the interstate. Kather was driving, slowly, mainly because she didn't have a driver's license, just a learner's permit, but I also think she was a bit preoccupied about crashing into a moose. The night before, a man at the hotel in Bangor had warned us about a rash of traffic fatalities on northern Maine roadways due to people crashing into moose—"the beasts are so tall that their bodies don't bounce off the hood or the bumpers like deer do; they smash right through the windshields and crush the people inside." To take her mind off the moose, or mooses, or meeses, whatever, I began talking about our years living in Korea. She was only four, five, and six back then. "Do you remember the traffic in Korea?"

"Not really. A little bit. I remember being the only round-eyed girl in my first grade class. My teacher was really prejudiced against Westerners."

"Well, there were fifty million people living in that country, and it's smaller than Maine. And yet everybody owned a car. It was like one big parking lot. You don't remember that?"

"Not really. I remember the marketplace at the bottom of the hill, and how all the men and women sat on their heels instead of standing or sitting."

---

* A character from day one, Kather started calling her parents "losers" at about age fourteen so as to not have to take them seriously. It's a term of endearment in our family.

"Yeah, well, believe me, it took an hour and a half to drive five miles across Seoul. So whenever I find myself driving on an empty stretch of highway like this one, I have this vivid image of a Korean man coming upon the same situation for the very first time in his life. He starts to speed up. He's giddy with road freedom. Cellos are playing in the background—"

"Cellos?"

"Yes, actually, it's an image with a musical score."

"What, are you gonna make a movie of it?"

"No, it's just a fantasy. Anyway, cellos are playing a slow melody as he continues to speed up. Pretty soon the pace is over-the-top fast, and the cellos are replaced by violas, and as he goes faster and faster, the violas are replaced by violins. Then he starts to swerve, losing control. And just before he drives off a cliff, the symphony swells into cacophony. There's a fiery explosion. Cymbals crash and fall. I call it my Korean Movement in D."

Kather didn't say anything for a time. Then she slowly looked over at me with a serious frown: "In *D?*"

"Yeah, D for *Drive*."

Another long silence. "It's a movement all right, but the D doesn't stand for *drive*."

As we came to the very end of I-95, which stretches two thousand miles from Maine to Florida, and continued north along Route 1, which runs parallel to and a mere five miles west of the Canadian border, we were amazed, and certainly a bit relieved, to find ourselves surrounded by farmland and roadside businesses. The wilderness that we had driven through during the entirety of the morning hours had disappeared completely. If I had been placed down there and not told where I was, I would have guessed central Wisconsin or Minnesota, not northern Maine. And as we rolled ever north, on both sides of the road, and on every farmer's field, they were harvesting potatoes. This was Aroostook County, Maine.

Because the country north of Bangor had been so wild and remote, we had not dared to get off the road to find something to eat. Suddenly we were in a heavily populated zone with cars and stoplights and pizzerias. Not but five miles from our destination, we came around a turn and there up ahead was a charming little diner called the Blue Moose. It had all the earmarks of a great little eatery. Great it was. Little it wasn't. Kather ordered a BLT, and I ordered chowder and a fish sandwich. What a heaping plate of food they brought us, and it was as good as it was ample. Best chowder I had had in a long time. And as we groaned with satiation and waddled sleepily to the car, Kather commented: "I sure hope they don't put us right to work because I'm not very good after a big meal like that."

We drove past a large field with a handful of cows staring blankly at us. We turned onto a dirt driveway and rolled slowly forward. There was a good deal of clutter along the way—machines mostly, a Quonset hut or two, piles of pipes and gadgetry, some chickens. There were thick woods on either side, but after a hundred yards we came to a large clearing of buildings and gardens and hoop houses and more clutter. It felt like an old-time settlement. The dominant building was set low to the ground and extended a good fifty yards or more. It was stained dark and looked like a windowless factory made of wood. Built perpendicularly and into the side of the windowless structure was a garage or warehouse that yawned at us, its wide-open mouth full of sacks of potatoes and shelving. Chickens ran loose everywhere. There was a plethora of automobiles parked next to the house, all fully functioning. There was one open spot, and we pulled in.

As we got out of the car, three dogs ambled up to greet us. One was a border collie mix, another a corgi, and the last a huge white Great Pyrenees. All three were a little wary of us, but after the border collie mix accepted our affection, the other two quickly befriended us.* Next we heard what sounded like a motorcycle growing closer. Suddenly a teenage boy appeared from around the building on a very loud ATV. He was a handsome, dark-haired, and very dirty lad, with a determined look on his face. He idled down as he came abreast of me, and I told him who I was. He motioned for us to follow the road around the house, in the direction he had come, to the field in back where they were harvesting. We did just that, followed by the three dogs.

Up ahead we could hear and see a large machine grinding out in the field. We passed a small apple orchard ringed by an electric fence with two huge black pigs behind it, eating apples and following us with their large heads and eyes and snouts. As we approached the work site, a dozen people came into view. They were all gathered at the edge of the dirt field watching the action. Jim Gerritsen, the farmer and owner of the Wood Prairie Farm, was pulling and prodding at a huge boulder in the dirt with a bulldozer. And just as we joined the group, the machine suddenly lurched forward and a fountain of dark hydraulic fluid sprayed into the air. The hose had sprung a leak. There was a

* Their names were Happy, Jackie, and Snowball, respectively.

great burst of shouts, all in unison, as the folks next to us alerted Jim of this unfortunate occurrence. He turned off the machine and hopped down to have a look. I took the opportunity to look around myself, and in every direction all I saw were rocks—piles and piles of them. And out beyond the dirt potato fields, perhaps ten acres or more that formed an L, was dark and wild forest, thick with pine.

Shaking his head and swiping his greasy hands together, Jim bounded over to welcome us. Tall, steel-eyed, fiftyish with an august mien, with a closely cropped beard, red flannel shirt, and brown corduroy pants, he was the spitting image of my favorite president, Abraham Lincoln. And he talked like him, too:

"It's the darndest thing, a rock that size swelling up so late in the year. Wasn't there last year or the year before that. They just keep a-bubblin' up from the center of the earth."

"Looks like you have more rocks than potatoes around here." I stated the obvious.

"You got a keen eye for details," he smiled.

He welcomed me to his farm and introduced Kather and me to his entire family. There were his two daughters, Sarah (thirteen) and Amy (eight), and the two older boys, Caleb (seventeen) and Peter (twenty). His wife, Megan, who was originally from my area of the country, near Syracuse, New York, was a soft-spoken, pretty woman with blond hair and the most muscular hands I'd ever seen. She looked like, and had the physique of, a former Olympic athlete. There was also a handful of teenagers from town, and a housewife in her forties, Robin, a recent empty-nester. All of them were wearing canvas gloves.

Aroostook County, Maine, is one of the last areas in the United States where schools are closed during harvest season so

that kids can help farmers get their crop in. Often the teenagers who are hired are taught potato-picking techniques by their parents and grandparents, who, in turn, learned from their parents and grandparents. Which all speaks to the fact that Maine has a large percentage of family-run and -operated farms, and at 5 percent, Maine has the second-highest ratio of organic farms to conventional farms in the United States, behind Wisconsin.

Jim gave his son Peter the task of driving up to "the city" of Presque Isle to get a replacement hose for the bulldozer, a good forty miles round-trip. Caleb, the boy on the ATV, was sent off in search of tools. Next Jim pointed to the huge red contraption sitting behind a large green tractor out in the middle of the rock-strewn field.

"You ready to do some potato harvesting?" he enthused.

I glanced at poor Kather, whose eyes got as wide as saucers, and whose hands went immediately to her belly. I reluctantly nodded. "Sure."

"Well, follow me."

He took us out to the middle of the potato field. There was a green tractor with a red harvesting machine behind it. We stood next it, looking back across the fields. I could see the rows that had been harvested, and the ones that were still mounded up, with the charred black plant tops sticking out.

"Why are the plant tops all charred?" I asked.

"I flame the tuber heads. It kills the germs so we're not spreading virus, which is particularly important to us, being seed sellers."

"Do you have a flamethrower or something?"

"Yes, there's a machine for that. First we strip the tops with a flail mower, then flame them. Chemical farmers top-kill with a chemical desiccate two to three weeks before harvesting. Not good stuff."

He then walked us around the harvester. Pulled through the potato patch by a tractor, this amazing machine has a blade that lifts the potatoes onto a metallic netting, where the loose dirt and rocks fall out. The potatoes, and rocks, then move back through the harvester onto rollers and then onto a picking table, where the crew picks out the stones and the clods of dirt. On top of the conveyor hierarchy there are a series of mechanized filters set at obtuse angles that are made of either steel or rubber that help separate out rocks and clumps of dirt. The conveyor lines go in two directions: One goes straight back, like a tail that sticks out from the harvester a good four feet. This is where extra-large stones and clods of dirt are excreted straight away. A single person sits on a welded-in-place seat and mans that station, helping to clear the line of any stuck stones or dirt. The main conveyor line that carries the potatoes turns 90 degrees and slants upward about 30 degrees, onto an elevator. Half a dozen workers man this area, or "picking table section," on either side. At the end of the line is a huge four-by-four-by-four-foot-square wooden box where the potatoes (and rocks) ultimately end up. Rocks that get separated out from the picking table go into a side trailer.

As for the potato pickers themselves, we were lined up, standing three per side. Our task was simple: remove rocks and toss them not much more than an arm's length into the trailer attached to the side of the harvester, which would be emptied every fifty yards or so, when full. One final person was perched on top of the wooden box where the potatoes ultimately fell, and plucked out any rocks that had managed to escape the twelve stabbing hands patrolling the picking line in front. It was pretty intimidat-

ing, all that metal-linked belting and all the metal edges around it. It would be pretty darn easy to lose a finger. I could tell Kather was particularly uneasy about it with her long and tapered bass-playing digits.* But engineers have a way of making things less dangerous than they could or would be otherwise. Plus, we all wore fat canvas gloves. "Soft hands," we were told.

And so it began. With Jim at the helm, seated behind the wheel of the tractor, he gave a great shout-out and the harvester began to churn. It was a deep, clanking, grinding, bass-noted drone. Then the conveyor commenced moving, making a metallic noise, reminiscent of a chain being dragged across a floor like something out of a Poe story. The tractor crawled slowly forward. It took a few moments, but eventually the potatoes began to appear, along with some rocks, on the conveyor. I was the last man at the end on the tractor side. It was my job to pluck rocks out, and if there were any large potatoes, I was to pull them out and set them aside in a special bin at my elbow. They would be placed in a different category for sale later on. Kather was on the other side of the conveyor, in the middle.

At first it was an occasional rock that came my way and an occasional big potato, but fifty yards down the line, rocks began to overwhelm us. The harvester rattled woefully and choked and barked as we struggled to remove the extra burden. And at the point when we couldn't keep up—the rocks being ten times more numerous than the potatoes—the first girl on my side of the line, a redheaded teenager with a solid neck and amazingly fast hands, hollered out, and Jim stopped immediately, idling

* Prior to her dropping out of college, she was the principal bassist for the University of Buffalo's symphony orchestra.

down the motor. We all gave a great sigh of relief and began to remove the stones.

It took us a few minutes to clean away all the stones and to empty the metal bin that held the rocks we had removed. There was a huge lever that was pulled, and a thousand pounds of stone dropped out, literally shaking the ground and leaving a large triangular pile there in the middle of the field. I soon noticed dozens of these triangular piles out in the field we were harvesting.

I just could not get over the concentration of rocks in the soil. In Kansas and in Michigan there was not a single rock anywhere in the dry, sandy soil. I was also awed by how slow and grinding the pace of the work was. Jim drove the tractor in low gear and sat sideways so that he could watch us pick and watch the row in front of him at the same time. Each row was about a football field in length, and it took about thirty minutes to go around from the start of one row to the start of the next one, depending on how many times we had to shut the operation down due to the overconcentration of rocks. Then, once we hit the end of the row at the edge of the forest, we would all jump off, grab a bucket, and walk back up the row, picking up those potatoes that the harvester had dug up but failed to consume. Meanwhile, Jim would circle back around with the tractor to the beginning of the next row.

As I stood with my bucket of stray potatoes, waiting for the tractor to swing into place, I counted down the rows we had left to harvest in that field—twenty-three, about twelve hours of work. And that was just half of one field. There were five more full fields left. Later I would learn that the Wood Prairie Farm grew eighteen different kinds of potatoes on about forty acres, ten acres per season. The row we were working was the Red Russet variety. There was a lot of work to do.

I looked over at Kather, who seemed a bit dazed by it all. I had been watching her on that first pass, and her work on the line was pretty slow and diffident. I was going to say something to her, but then out of nowhere, as if she had been reading our minds, the girl who had been manning the tail section clearing stones and dirt clods off the back of the harvester wanted to know if Kather would like to switch jobs. Kather brightened immediately, and the girl handed her the tools of the trade—a mason's trowel and a protective face mask like the kind carpenters use when working a lathe or a ripsaw. Kather assumed her seat at the tail section, and that would be her post for the next three days of the harvest.

Whenever we came to the end of a row, the entire gang would jump off the tractor and romp with alacrity across the field, hopping, skipping, and gleaning potatoes. The dirt was so soft and rich that you felt like you were sinking down in it as you trudged across. Often, Sarah and Amy and some of the other high school kids would perform gymnastics, doing back handsprings and cartwheels in the loamy soil. It certainly set the mood. There was such a sense of purpose and team spirit to this work, as loud and as mechanized and as monotonous as it first appeared to be, that every face sported a contented smile. This was a magical place, this organic potato farm in rural Maine, run and operated by a man and his wife and family. As Jim himself put it, "The essence, the real value of life, comes from what we're doing, the work of our lives, the people we're interacting with, and it doesn't come from the amassing of goods."

We worked until six o'clock that evening, which was about four hours for Kather and me. As we left the area, Caleb came along driving another tractor; he was pulling behind it an odd machine with chain-link netting, five feet wide, that revolved like the conveyor on the harvester, only the chain link sagged in the middle. I watched as he entered the field we had just left. He dropped the front down and began to harvest the rocks in the field, avoiding the triangular piles, which, I was later told, would be gathered with a front loader anon. It was a rock remover! Meanwhile, Jim had hopped onto a forklift and hoisted up one of the full potato boxes. We followed him back to the house, the dogs at our heels. We stood and watched him and Peter as they loaded three full four-by-four crates into the warehouse that was attached to their home. It was an interesting setup. Thirty feet to the left of the front doorway of the home was a garage door that sat in the middle of a long, windowless section of the building. Opened, it revealed a vast warehouse eighty or ninety feet long and forty or fifty feet wide, and sunk down into the ground a good ten feet below ground, with thick cement walls up to the ground level, and wood siding from there up to the ceiling about thirty feet above the slab. Fluorescent lights illuminated the interior. There were crates stacked on all four sides, with bright labels on them. Jim, once he'd opened the garage door, disappeared inside, and he was now driving another forklift down in the hole as we stood at the edge of the steep, ladderless drop-off watching the action inside.

We watched as he drove the forklift over toward us, picked up the first crate from the edge of the drop-off, did a complete 180, and drove it over to a spot along the far wall and stacked it. Both father and son are expert forklift drivers. Later I learned that during the harvest season they will stack and store somewhere in the neighborhood of two hundred thousand pounds of organic seed potatoes in the warehouse. The crop overwinters there. It is usually all sold off by the Fourth of July. The potatoes that don't get sold are eaten by the family during the summer until the next crop is harvested. The potato salad they make from the storage potatoes is a staple of the Gerritsens' summer meals. Talk about stability through good eating!

We were invited to dinner. Peter, in his task to get a new part for the bulldozer up in Presque Isle, had also bought a month's supply of pizzas. There were boxes of pizza pies stacked up all over the kitchen. Now, I am certainly not the kind of person who would be critical or judgmental of another person's home. In my book, every home is a castle. In fact, I can truthfully say that I feel more comfortable in a filthy, cockroach-infested trailer, home to a dozen Mexican field hands, than a multimillion-dollar oceanfront summer estate kept cleaned and polished by a team of domestics.* So please believe me when I tell you that this home was like none I had ever experienced before in my life. When I say there were pizza boxes stacked all over the kitchen, I am not exaggerating. The reason they were all over the place was that there was not a single surface available to put them on.

---

* After graduating from college and hiking in the Maine woods, I traveled across the United States, staying and working at different places around the country. One of the first places I stayed at was a friend's rich great-aunt's august manor on Martha's Vineyard, in Edgartown. Soon after, I found a job working on a two-thousand-acre hacienda in Texas as a ranch hand. My roommate, Acedro, would take me along with him into Houston on the weekends to visit his uncle and cousins, who lived all crammed and jammed together in a little trailer. It was the dirtiest and most delightful home I've ever slept in. Guess which place I liked better?

Even the kitchen table was log-jammed from end to end with indescribable stuff. And this stuff, this clutter, these soaring piles of nondescript items, which seemed to swirl up and out of control like an infestation of locusts, onto the walls and ceiling, was one hundred percent objets d'art*less*! There was not a single piece of functionless beauty or precious eye candy or attractive bric-a-brac to be found. These items were like cells and organelles and organs of a vast and complex organism. It made me think immediately of the late, great British travel writer Bruce Chatwin, who wrote copiously about the immorality of *things* and how they have an uncanny ability to insinuate themselves, with disastrous effects, into people's lives.* These potato people were truly the *masters of their domain*.

You entered the kitchen through a mudroom, which was the most literal example of that word I'd ever had the good fortune to experience. Boots and potatoes and *sundry* were legion there. The next door led you into the kitchen proper. To your right as you came in was Jim's computer and desk area. You could hardly see the computer for the stuff. Now Jim, as president of various organizations and owner of a seed company that puts out a catalog, is an accomplished writer who is constantly at work on some newsletter or other when he is not out in the field planting or harvesting. Somehow he manages to get things done at that

* Perhaps the quote I was thinking of was the one from "The Morality of Things" that reads: "But things have a way of insinuating themselves into all human lives. Some people attract more things than others, but no people, however mobile, is *thingless*. A chimpanzee uses sticks and stones as tools, but he does not keep possessions. Man does. And the things to which he becomes attached do not serve any useful function." Bruce Chatwin, *Anatomy of Restlessness: Selected Writings 1969–1989* (New York: Penguin Books, 1996), pp. 170–71.

workstation. Two steps farther, beyond two sturdy beams, you are right there in the kitchen. Pots and pans and dishes erupt out of the two deep sinks. There's a fridge somewhere over there next to it, as well as a stove and various kitchen tools. Dirt from the fields is part of the décor. If potatoes were people, this is how and where they would live. Later Kather put it perfectly to words: "They are outside people, Dad. Inside is just an extension of outside."

To the right of the kitchen proper was a little sitting area with a long, high-backed couch with countless cats lounging on top of it, and a large television sitting on top of an old piano. There were shelves of books around, but even the books were in disorder and clothed in potato dirt. To the right of the television and piano was a tiny, narrow staircase, almost like one you'd find in a tree house, no wider than a grown man's shoulders. And the door at its terminus was equally small. But the tiny, narrow stairs bifurcated, turning 90 degrees to the right and disappearing at a steep angle up into darkness. It struck me that the Gerritsens lived like the Swiss Family Robinson, Robinson Crusoe, and Jean-Jacques Rousseau all rolled into one, living the lessons of frugality, husbandry, and natural history each and every day. But as I soaked it all in, I wondered, where would Kather and I sleep?

"So, what are your sleeping arrangements?" Megan suddenly asked me.

"I'm not sure. We'll go into town and find a motel," I said.

"Well, I think I mentioned in my e-mail that we have a cabin out back, and you're welcome to stay there if you'd like."

"That sounds great."

Next thing we knew, Kather and I, along with Sarah and Amy and Megan and the three dogs, were walking into the dark woods, a quarter mile from the house, along an overgrown path. The cabin suddenly popped into view the moment we stood in front of it, hidden under the pines. It looked like the top section of a house, or a dormer, that had been torn off during a storm and tossed there on its side, in the gloaming of the woods. It made me want to look and see if the feet of the Wicked Witch of the East were sticking out from underneath. Megan and the girls were all happy and excited about visiting their old playhouse and burst through the windowed entrance door with shouts of glee.

"This is where we first lived. Well, it's part of the building we first built and lived in," Megan informed as we entered inside.

It was dirty and dusty, and full of spiders and bugs and cobwebs. There was a wood-burning stove in the middle of the room with candles on it. There was no furniture to speak of, just wooden floors and boxes of things here and there. No running water, no electricity, but, oddly, there was a television set, even though there was nothing to plug it into. There were windows, some cracked and broken, looking out at the trunks of moss-covered trees behind which lurked lions and tigers and bears, *oh my*.

"The loft has mattresses," Amy, the youngest, happily informed us as she pulled down the little wooden ladder to the loft, climbed up it, and lifted open the trapdoor, showing us the loft and the two mattresses lying on the floor. The loft was three feet from floor to ceiling, no more. The mattresses were foam rubber and full of dust and spiders.

"Let me show you the outhouse," Megan suggested.

"Good idea," I enthused.

We followed her outside. The outhouse was right next to the

cabin, just as crusted and infested with spiders, bugs, and cobwebs as the cabin.

Again, my patient reader, these descriptions I give are not meant to be taken in any way as complaint or derision; rather they are to be considered as an admission of guilt from an aged hypocrite who might have preferred the toilet-clogged miasma of a trailer-full of Mexican farmhands at age twenty-two, but who at fifty-two much preferred the queen-size mattresses, glistening clean porcelain, and hot running water of modern-day hostelry. I smiled and nodded, but the moment I could pull Kather aside, I informed her:

"We're not staying back there. We're going to a hotel."

"You're a wimp, loser," she said, frowning. I could tell she was disappointed. Of course she was; she was twenty-two.

"Yeah, well, that may be, but it's out of my comfort zone."

"Comfort zone? I didn't know you had one."

"Oh, I've got one."

"You just can't crap in an outhouse, that's why."

"Yeah, well, for me, outhouses have a bad association."

"What kind of bad association?"

"I mean literally a bad association: the Outhouse Guild is nothing but a bunch of finks, crooks, and inconsiderate slobs."

"Is that the best you've got?"

The sun was low in the sky as we made our way back to the house. Inside, we ate slices of pizza and sat at the kitchen table, talking with Jim and Megan for hours while the kids sat and ate pizza and watched television and eventually slipped away, back into the warrens of the tree house. In truth, Kather and I said very little. Jim did most of the talking. He spoke grandiloquently and didactically about the history of the area, the logging, the potato farms and farmers, and ultimately about organic farming and farming in general. It was edifying, to say the least. Pieced together from notes I scribbled and lines I copied from transcripts of talks he has given, here's roughly what he said:

"The loggers who came to the area back in the early 1800s soon discovered that the well-drained, rocky loam along with the cold weather were perfect for growing potatoes. It took about one hundred years, but the Aroostook County farmers slowly and steadily cleared the forests from hundreds of thousands of acres in order to grow potatoes. Then when the railroads came through in the late 1800s, that cinched the deal. Maine was the potato capital of the United States. Even today Aroostook County produces more potatoes than any other county in the United States. And yet Maine is no longer the Potato King. This is the result of a number of factors. The first is the shifting consumer preferences away from fresh potatoes and home-cooked meals to factory-processed foods, such as McDonald's french fries and Lays potato chips and Swanson's frozen french fries. Another reason is that the potato growers of the American West have an unfair advantage: they benefit from federally subsidized irrigation and hydroelectric projects. You add to that the transformation of the traditional potato to a generic commodity with a capital-intensive, highly mechanized, increasingly concentrated system of large-scale, factory-like production. And finally, decade after decade of low farm-gate prices. We've had a tough time up here.

"As far as the Maine organic community . . . our ten acres of potato production would be considered average. However, it is very small compared to the 200- to 700-, even 2,500-acre potato crops of our nonorganic neighbors. Yet the size of the thousands of farms back in Aroostook's golden age were also small in comparison to

today's. As average farm size increases, there is an even greater decrease in the number of farmers remaining. Fifty years ago our four-mile-long stretch of road here had thirty potato farms. Twenty years later our entire town was down to thirty potato farms. Today there are just six potato farmers left in town. One economist has projected that if current trends continue, Maine's sixty thousand acres of potatoes will one day be grown by just twenty farmers each growing three thousand acres. This upward trend in scale is the same across American agriculture. Two factors contributing to this trend are the short-sighted acceptance of genetically modified crops by American farmers and the unrelenting rise of American corporate consolidation and domination, first within the U.S. economy and now within the national government.

"Since the beginning, the organic community has been a safe harbor for the American family farmer. But organic family farmers could easily succumb to the same forces of scale, consolidation, and control that have led to the demise of other family operations. There are three reasons for hope, however. The first one is the spirit embodied in movements like Slow Food that honor the producer and value food that is good, clean, and fair. Second is the developing concept of an 'Organic Family Farmer' certification system that identifies organic family farmers in the marketplace, aiding coproducers seeking authentic goods and protecting real family farmers from corporate imitators. And third is the dawning on Americans that local is in fact better, as reflected in the growth of CSAs and farmers' markets."*

---

* Much of what you read here is taken from the transcript of a talk Jim Gerritsen gave at the Slow Food convention called Terra Madre, in Turin, Italy, on October 28, 2006.

"So then which of the three is the most important when it comes to buying food? Organic, local, or family owned?" Kather asked.

"All of them," Jim replied immediately, almost as if he was insulted by the question.

"Yeah, but you don't always get that choice. Like where I live in Buffalo, there is a local farmers' market, and they do have produce from a couple of local farms, but the local farms aren't organic or family-owned and -run. But there's also produce from family-owned, organic farms that aren't local."

Jim had to think about that one for a second. "Hmm. Well, organic is first. That's the most important. Then family-owned, and then local."

It was getting dark out by the time we left. Lying that I hadn't taken a shower in days due to my travels, I thanked them for their generous offer of the cabin but told them that I preferred to find a room up in town. They didn't seem insulted in the least, and in fact gave me the name of a motel. The plan was to start early the next morning, at seven thirty. However, if it was raining or wet, they would have to postpone harvest until things dried a bit. He took my cell phone number and said he'd call and let me know some time before 6:30 a.m.

Out in the car, Kather said, "God, you're a terrible liar."

"You noticed that, huh?"

We drove north about five miles and came into the hamlet of Mars Hill. We found a neat little motel, the very same they had suggested, which had a vacancy and a room with two queen-size beds and the brightest lighting I'd ever experienced outside of a hospital operating room. The sun had set and we could just make out the town's namesake across the street, Mars Hill itself.

A miniature Vesuvius—it's not even two thousand feet—it kisses the border between the United States and New Brunswick, Canada. We were told that it's the first place in the United States to see the rays of the rising sun. There were giant windmills on top, spinning lazily.

I hit the sack and was out almost immediately. The phone did not ring, because I had forgotten to turn it on. As a result, we got up late and had to scramble to get back to the farm by seven thirty. It was drizzling and misty again. When we got there, Jim was inside at his workstation, typing a blue streak. When he realized we were there, he came and stood outside with us, scratching his neck.

"I tried calling you. There was no answer."

"Yeah, sorry, I forgot to turn my cell phone on. I don't use it very often, so I'm not in the habit of turning it on or off." I felt obliged to explain my stupidity.

"Well, it's too wet to dig potatoes. We've been pushing pretty hard, digging every day up through last Sunday night, and since then we've been too wet to dig most mornings. We're not getting the dry, hot falls around here anymore. We've had this same wet weather during harvest season the past four years. Kids go back to school Tuesday and it will be nip and tuck to get the potatoes out by that time. Heavy, wet soil is hard for us to slog through—takes more effort and production slows down, but, after all, we're survivors. Well, I have some catalog business that needs attending to. We'll try again today around noon. I'll see you then."

Without anywhere else to go, Kather and I returned to the hotel, and once there, with Mars Hill staring at us, we decided to go ahead and climb it. But when we got there, our wills had abandoned us. We leaned against the car for a long time, lazing around the parking lot of the ski resort there and saying nothing. Finally I had to ask: "So what do you think of the life of a potato farmer?"

"I think the way they live, with the dirt and the family and the community all together, it's like something out of a National Geographic special. They're amazing people, but you know what I'd like to do?"

"No, what?"

"I'd like to hide a Mr. and a Mrs. Potato Head out in the rows, so they get harvested."

"What?"

"We've got to find a store and buy a Mr. and Mrs. Potato Head, sneak out into the field, and hide them in the unharvested rows."

The idea suddenly sunk in. I began to laugh. "Can you imagine their faces when Mr. Potato Head hits the picking line?"

"Whose faces? The pickers or the Potato Heads?"

"The pickers."

"Yeah, but what about the Potato Heads? They have all those choices of faces—the different eyes and mouths and noses. Which faces should we put on them? The mean eyebrows or the surprised eyebrows, the big red lips or the shiny white teeth? I can't decide. Come on, loser, let's see if we can buy some!" She was quite inspired by the idea.

Unfortunately, we were unable to locate a store that carried the Mr. and Mrs. Potato Head line of products. You'd think that in Aroostook County every store would carry Mr. and Mrs. Potato Head. We were pretty down in the mouth about it, too, especially Kather. So we assuaged our disappointment by returning

to the Blue Moose for an early lunch. I just got a big bowl of chowder and shared Kather's heaping pile of fries as well as a bite of her BLT repeat. Before we left, I told the owner that I'd just come from Boston and had had a bowl of chowder there that didn't hold a candle to his. He thanked me profusely and said that they didn't know how to make chowder anymore; they used too much roux. I agreed.

When we got back to the farm, Jim was outside finishing up his repair on the tractor's hydraulic hose. He wiped his hands on a dirty rag, came over, and told us it was still a little too wet to harvest; he wanted to wait another hour. None of the crew was there yet anyway, so he took the opportunity to give us a tour of his farm. And with the dogs following along, he told us how he grew his potatoes and how he ran his business and why he loved farming. Most of what he said I understood, some of it I didn't:

"I came here in 1976 when I was twenty-one. An acre of land was selling for a hundred and fifty dollars. We have a hundred and fifteen acres in total, with fifty-six acres of it in farmland and the rest in woods. Our first business was cider. Remember the apple orchard with the two pigs that you asked about? Our second plan was a CSA. But by the early nineties, we were fully into heirloom seed potatoes. We farm about fifty-five acres, forty-eight acres of it in rotated crop production. It's a four-year rotation cycle. The first year is potatoes. The second year is spring wheat or oats. The third year is clover and timothy grass. The clover sod from the third year is plowed down in year four and the field is then planted first to plowdown buckwheat, then to plowdown rapeseed as a bio-fumigant.* The fifth year is back to potatoes. The rotation cycle produces about ten to twelve acres of potatoes a year. We also grow some other root crops, like carrots, beets, parsnips, and onions. We plant in May, harvest by early October, and ship from underground storage until the Fourth of July. We sell seed potatoes certified as seed by the State of Maine to home and market gardeners across the United States through a mail-order catalog and website. We also wholesale to mail-order seed houses that sell our organic seed potatoes in their catalogs."†

As we walked, he showed us the two man-made ponds that he and Megan had dug by themselves years before. They used them to irrigate the fields, although lately it was so wet they didn't need them. The smaller pond was close to the house. It was about fifty by eighty feet, and all around the edges there were canoes and kayaks and floating devices that the kids played with. But at the back section of his property, ten feet above the potato fields and the road, lay his magnum opus. It was an enormous pond, three times the size of the other. There were fish in it. It was deep, too. This was Jim's hedge against drought. He had built it years ago, dug it himself with a backhoe he rented. "Cost me a king's ransom in fuel," he said. The steep sides were riprapped with boulders and lined with clay. The forest pressed in

---

* Grain crops like buckwheat and rapeseed are most often grown for grain harvest. Organic potato farmers like the Gerritsens use these two crops for *plowdown* at optimal growth for the soil; hence, they are called *plowdown* crops.

† Parts of this conversation were grafted from the same speech Jim gave at the Slow Food convention at Terra Madre in Turin, Italy, on October 28, 2006.

close on two sides; fields and road accentuated its height on the other two sides. I could tell he was still a bit in awe of his own youthful energies to have accomplished such an engineering feat. And it was truly a marvel in relation to this small-scale operation. He had lifted an entire acre of land ten feet up into the air and dug a huge hole down into it, filled it with water, and stuck pipes and pumps in and around it. We stood on the tall rim, looking back down at the potato fields and the orchard and the buildings. This was his home and land. He had created it with the sweat of his brow and his own two hands. As an itinerant, I could only imagine his profound sense of place. I took a step closer to him and slipped invisibly into his shoes. I stood in them feeling the active root systems stretching forth from his toes into the soil, pumping up nutrients and exuding rare earth elements. There's a world of difference between a place to hang your hat and a place built from your dreams.

The sun had come out; evaporation was in overdrive. It was time to harvest again.

And we did so until five thirty that evening. It was the same crew as the day before, except for Robin, who had a part-time job in town and couldn't work in the afternoon. Kather assumed her spot at the tail end of the harvester with her face mask and her heavy-duty trowel. I only saw her occasionally on my way back across the field, gleaning potatoes. She had befriended one of the girls in the picking line and would walk and glean with her a lot of the time. I was in the same position, too, plucking out rocks and thinking constantly about Mr. and Mrs. Potato Head showing up on the picking line, their eyebrows bent at odd angles, their eyes and red lips and white teeth caked with dirt . . .

At the end of the day, on the long, slow walk back to the

house, Megan and Amy approached me and Kather. Megan asked Kather if she would like to milk a cow. I knew for a fact that Kather had always wanted to milk a cow, so within minutes, all four of us were piled in the cluttered Dodge van and driving out the driveway and down the road toward a large field where four cows stood, chewing the cud alongside an electrified fence. One of the cows, the tallest of the herd, was a black and white Holstein. I recognized that breed. He was friendly and came over to greet us immediately. I rubbed his head and scratched the soft area around his larynx. But the other three were spooky. They did not come over to greet us. On the contrary, they sort of skittered away. They looked Asiatic, and I asked the name of the breed. "Dexter," she said. They were as broad as they were tall, and dark black. Megan told us that they were a very maternal breed of cow, great for milking, but of difficult temperament.

There was a small trailer parked on the edge of the field where Megan did her milking. She opened the back door and set her equipment inside, jars and buckets, and then proceeded to chase down the one cow she was there to milk, pulling a collar over the largest female and dragging her into the milking stall.

She set about milking her. It wasn't easy. The space was kind of small; the cow kept jerking and trying to back out. When she was fairly settled, Megan called Kather over to give it a try. I was sure impressed with my daughter bounding right up there and slapping those elongated paws around those teats and yanking away. She actually did it correctly, too. She had a big old grin on her face when she was done.

"Good job, loser," I said with a nod as she came over and stood beside me.

"It was a lot easier than I thought."

"So do you folks sell the milk or drink it yourselves?" I asked Megan.

"It's just for our table. We drink raw milk."

"Another raw milk drinker. Almost all the farms I've worked on have drunk raw milk."

"It's the best," she said matter-of-factly.

"I'm beginning to believe you."

They fed us salad and homemade bread and raw milk for dinner. The whole family was there. We ate together. Again we spoke for hours to Jim and Megan, same as the night before. Early in the conversation, Megan, with the kids all around and listening in, asked Kather about stand-up comedy. Then she wanted her to tell a joke. Kather politely explained that she didn't work that way, but followed up by telling her that comedians were like farmers, because they are the only people telling the truth these days.

That led to the next topic: the case against Monsanto. Jim began by talking about Bt.

"Do you know about *Bacillus thuringiensis*, or Bt?"

Both Kather and I responded, "No."

So he explained it flat out, in his didactic style, his cadence and exigence sounding at times a bit like a Russian revolutionary's.

"The bacteria Bt grows naturally in forest soils. Some subspecies have been commercialized. Bti [*Bacillus thuringiensis israelensis*] controls mosquitoes. Btk [*Bacillus thuringiensis kurstaki*] controls lepidoptera [larvae of moth and butterfly] and is widely used by organic farmers to control crop pests. Bt is very important to organic farmers and is commonly used by them. The stories you've heard recently about pests evolving resistance to transgenic Bt made by Monsanto are true, and that's a big problem for us organic farmers and gardeners because it's one of the only natural, nonchemical pesticides we have available to us. Because Monsanto has created these transgenic crops that have the Bt gene in it, like any overuse of herbicide or pesticide, the insects grow resistant to it.

"In the mid-1990s Monsanto introduced their 'New Leaf' potatoes, which were gene-spliced with Btt, the Bt subspecies [*Bacillus thuringiensis tenebrionis*] that is effective at controlling Colorado potato beetle. For many years, when the Colorado potato beetle pressure was high, organic farmers and gardeners would spray Btt on the potato plants in the late afternoon. Overnight the beetles would eat the leaves and ingest a lethal dose of the Btt, which would immediately paralyze their gut and kill them within a day or two. As soon as the sun's rays got to shining the next morning, the Btt would break down and be gone within a day or two of application.

"Normally, depending on how bad the beetle infestation, we would spray Btt a second time, again in the late afternoon, about seven days later in order to control larva that had hatched subsequent to our first Btt application. Again, kill took place overnight and again the Btt began to break down in the presence of sunlight that next morning. What's important to understand is that insect resistance to a toxin develops in direct correlation to length of exposure (more exposure equals faster development of resistance). We were grateful that Btt did its control job quickly and then disappeared.

"A typical potato variety grows for one hundred and twenty days. In our organic potato field, for example, Colorado potato

beetles typically interacted with the Btt for just ninety-six hours, or four days out of the hundred and twenty days, or 3.3 percent of the season. Because Monsanto had gene-spliced the Btt toxin into their New Leaf potatoes, every cell of those New Leaf potato plants, including the tubers, which people eat, expressed that bacterial toxin. Over the short term the New Leafs controlled the beetles. But you have to understand that the New Leaf version of that hundred-and-twenty-day potato variety expressed the Btt for the entire hundred and twenty days. That means the beetles interacted with the transgenic Btt all one hundred and twenty days, or one hundred percent of the season.

"Resistance was absolutely predictable. In fact, in April 1996 I attended a USDA conference on transgenic Bt resistance held in Bethesda, Maryland. The nation's leading entomologists were also in attendance. Their clear consensus was that transgenic Bt resistance, whether in potato [New Leaf was rejected by the market and removed from production around 2001], corn, or cotton, was not a matter of whether, but a matter of when. So transgenic Bt resistance was predicted fifteen years ago. No one should claim surprise with new reports of transgenic Bt resistance.

"Once a population of insects becomes resistant to a material such as Bt, there is no going back: the material will never again work as a control. It is ruined for all of us. It seems that justice would require culpability for an entity that misuses a treasure taken from the commons, like Bt, and ruins it. And most especially after they were warned at the start that their technology would result in this loss. This is where we're at with this case and this arrogant corporation. They are claiming that their product is improving agriculture when in fact it is just the opposite; it's doing irrevocable damage.

"We have multigenerational farms up here on the brink of going out of business, and the last thing you want to do as a multigenerational farmer is lose the family farm. But if Monsanto wins, all we'll have to eat in this nation is GMO produce, and most of it is grown overseas. That's not a healthy food system. The whole point of organic farming is the idea that the crops are pure and with nothing transgenic or modified in them, but once our organic seeds are contaminated, there is no way back. One thing is certain, though: the end result will be that Monsanto either controls the seeds or has contaminated the seeds. No one wins except Monsanto."*

Kather and I sat there numb and silent for a long time, letting it all sink in.

"This is one of the most important issues of our time," Kather said, sort of to herself but out loud.

---

* At the time of the writing of this paragraph, Jim Gerritsen was named by the *Utne Reader* as one of twenty-five visionaries who are changing the world. Here's the announcement:

> ***Longtime potato farmer selected for efforts to protect family farmers***
> **Bridgewater, Maine**—Jim Gerritsen, a Maine organic potato farmer with a decades-long record of community involvement and activism, has been named by the editors of *Utne Reader* to the magazine's 2011 list of 25 "People Who Are Changing the World." Gerritsen was selected for his ongoing work leading efforts by independent family farmers to protect themselves from the threat of Monsanto litigation related to the corporation's patents on genetically modified seeds, an effort he sees as critical to the preservation of organic farming itself and organic foods as a choice for consumers and their families. Each year, *Utne Reader* selects 25 people "who possess an inspiring combination of imagination, determination and energy," said *Utne Reader*'s editor-in-chief David Schimke in a statement. "These are people who don't just think out loud, but who walk their talk on a daily basis."

I was gnashing my teeth, I was so mad: "This stuff makes me think of *Nineteen Eighty-Four* and the Party slogan written in giant letters on the walls of the Ministry of Truth:

WAR IS PEACE

FREEDOM IS SLAVERY

IGNORANCE IS STRENGTH"

And Kather added, "Yeah, and it makes me think of *South Park*, when they're at the genetic engineering ranch and complaining that the government is abusing the Prehistoric Ice Man, and the federal agent turns to Kyle and says—'Little boy, sometimes what's right isn't as important as what's profitable.'"

Kather and I drove back to the hotel in silence. Just as we pulled in to the place, she said to me: "I want to be a farmer."

"What about comedy?"

"I can do both."

"You sure can."

That night, as we were lying on our beds at the hotel, it was as if Somebody Up There was having some fun with us. Inadvertently, we found ourselves watching a grotesque biography of a reality television star. Neither of us recognized this woman. Perhaps because she had reinvented herself at least a dozen times—body, mind, and spirit—there was nothing real or natural left to recognize. She was like the human version of a genetically modified crop. Everything about her had been re-created and reprocessed as a kind of marketing tool or promotional concept, with her as the item of consumption. She was an actress, a gourmet chef, an inventor, a film producer, a product endorser, a holistic guru, a weight loss advocate, a magazine columnist, an author, a high-powered advertising executive, and now a reality television star on, of all things, *The Real Housewives of New York City*. Oh yeah, and she was a mother, too. Kather and I lay there in open-mouthed disbelief that we were watching something and someone so paradoxical to our profound Wood Prairie Farm people.

When it was over, we turned off the television, faced each other on our beds, and processed it.

"What a perfect representation of a person who has no understanding of herself or real life. She is not even real. Certainly her face and boobs aren't her own. She is nothing more than a product to be sold and consumed," I said. "She's like a Lays potato chip."

"Yeah, while the potato farmer is the perfect representation of a person who knows himself and is in tune with himself, his land, his family, the world. He's not a product that we buy; he is a person we believe in and feel connected to. He gives us hope. The life of the farmer is the life of our salvation. He lives simply and feels deeply for all that surrounds him, from his family and friends and neighbors, to the animals and plants he keeps, the ground, the rocks, the earth beneath his feet. He does not need us, but we need him. I believe in the farmer. I have faith in the farmer. I don't believe a word that lady said or represents. Nothing. And certainly not reality television," Kather opined. It was a beautiful soliloquy. And when I asked her a question a few minutes later, she was asleep.

The next morning we arrived at 7:30 a.m. again, but work did not start until ten thirty. We ate pancakes and doughnuts and drank tea. Jim and I sat and talked for hours, until work started again. We had a lunch of corn chowder, fresh bread, and farm-fresh boiled eggs. We worked until five that evening. We wanted to leave a little early so that we would be out of moose mayhem range before it got dark.

Upon departure, Megan looked at me and said, "I'm very glad you two came and spent time with us, to talk about authenticity, because we only live in authenticity and have no contact and no concept of anything inauthentic."

"Well, I am very glad we spent time with you, because the world out there is saturated with the inauthentic, and without places like your farm, we all might lose our way," I said.

INTERLUDE

Slow Money, My Money, and San Francisco

FORT MASON, SAN FRANCISCO

OCTOBER 2011

Certain cities have magical effects on some people. Of course it all depends on how much moving around people do. I believe the more cities you visit, the more they affect you. As for the magic, it is the result of four factors: weather, geography, architecture, and people. The combinations and degree of extraordinariness of each factor create the magic. San Francisco is a magical city with extraordinary degrees of all four factors. I am strongly affected by San Francisco, its primary effect on me being eternal youth. I have only to walk down the street in North Beach and I feel like a teenager again. Heck, I don't feel like one, I *am* one.

I arrived in San Francisco somewhere around the middle of October for my final harvesting adventures. I had a full itinerary, too. First stop was the Slow Money Conference down at Fort Mason, across from Alcatraz Island. The topic was "Investing as if Food, Farms, and Fertility Mattered," and Wes Jackson was the keynote speaker. I very much hoped to meet him and to ask him a few questions. From there I would drive northeast about an hour to Winters, California, to Sierra Orchards, where I would spend three days harvesting walnuts. Then from Winters I would drive back west to Glen Ellen in the Sonoma Valley, for my final harvest—grapes. With eternal youth on my side, and brimming with cockeyed optimism, I checked into one of the world's great hostels—the Green Tortoise, on Broadway in the North Beach section of the city, right across the street from the literary landmark City Lights Books. And best of all, they had a lower berth available!

The Green Tortoise is one hundred percent pure grooviness. Located in the upper floors of two adjacent buildings constructed after the 1906 earthquake, it is psychedelic with flower power on top. You enter from bustling Broadway with strip joints (in good taste) all around and climb a steep flight of wooden stairs. There's that old-time hotel odor to it. At the top you enter the foyer, and on your immediate right is the L-shaped reception desk, with pigeonhole mailboxes behind it. There are usually two receptionists on duty, always pleasant and always happy to see you. To the left is the grand parlor, now a huge open dining hall, set under twenty-five-foot ceilings, with a stage on one end that looks out over Broadway for impromptu musical performances, and a full-size kitchen on the other end. You can refrigerate and cook your own food, although half the time the meals are free, so why even bother. Back beyond the reception area is the Internet room, with a dozen computers at every point along three walls—yours to use with round-the-clock free Internet access. Upstairs are the bunkrooms. And all for twenty-nine dollars a night!

After a shower and a quick check of e-mail, I went directly over to the bookstore to see if I could find one of America's great booksellers, Paul Yamazaki, at City Lights Books. City Lights is the Rosa Parks of independent bookstores, having drawn a line in the sand by publishing and selling banned books and only affordable paperbacks long about fifty years ago. Out on the street, in front of the store, with all manner of humanity traipsing by, Paul and I reprised our role from my last travel narrative and shot the breeze about past and future trends.* We talked mostly about agriculture and how we were both convinced that one day the great stories, the great books, the great buildings, the great authors, and great architects would take a backseat to the great farms and great farmers and great heirloom crops. And just before we parted, he slipped in a shiny note of optimism about the book publishing business, revealing that City Lights Books had had a record sales year.

Whenever I am in San Francisco, I love to go up to Telegraph Hill and indulge in a sort of visual heavy petting of the city. The lush, tree-topped bulge of land between the Embarcadero and the commercial downtown offers a perfect place to do that. You can see everything from up there. Standing on the steps below the iconic Coit Tower (just Google that name), looking out over the voluptuous San Francisco Bay, the two red spires of the Golden Gate Bridge poking through cotton-white ground-level clouds, the heavy brown brow of Mount Tamalpais in the background, it is always so clear to me why a young Jack Kerouac and his youth-crazed ilk would repeatedly zigzag back and forth across the manifest breadth of this nation until there was no youth left in any of them. Because they yearned to lick

---

* To read about our first meeting, check out *Seeds: One Man's Serendipitous Journey to Find the Trees That Inspired Famous American Writers from Faulkner to Kerouac, Welty to Wharton* (New York: Harper Perennial, 2011), pp. 98–99.

and grope and press their noses into the copious cleavage of the City by the Bay, much like I was doing.

That night, after dinner, I met up with an old friend whom I hadn't seen in more than thirty years. He and I had grown up together on the East Coast and had attended public school together, from kindergarten all the way through high school. Over the years we kept in semi-communication through our mothers, who were friends. We met outside the City Lights bookstore and barhopped from there. He introduced me to a few new joints in the area, but we ended up at our mutually favorite spot—where the live music is always free, and the clientele is always happy to see you—the Saloon off Grant.* We didn't do much reminiscing; it was too long ago to even bother with any of that. He had two boys about the same age as my two girls, late teens and early twenties, and that was the main topic of our conversation. However, he did manage to describe the trajectory of his life since last we'd been together and how he had ended up in San Francisco. After college, he became a bond trader at one of the large global financial services firms and sold his bonds from New York to Los Angeles until he eventually landed in San Francisco. Money was his raison d'être; he even admitted it. And he'd made so much of it that he was retired before he had turned

* "There can be little doubt that The Saloon, at 1232 Grant Street in North Beach, is one of the top blues bars in the City. Make that the state. Make that the world. Make the scene with music every night with two bands on Fridays and weekends. The talent runs the gamut from raw to top notch, always spontaneous, always lively. No tired old big names here." The preceding announcement can be found on their website: http://www.sfblues.net/Saloon.html.

fifty. It was uncomfortable, but there was not a single point in the entire cosmology of his career to which I felt compelled to pose a question or offer a comment. Nothing! I did a lot of nodding. Likewise, he was not in the least bit interested in what I was doing in the Bay Area, or the books I'd written or what kind of life I was living. He nodded a lot, too, and asked virtually no questions. It was a head-scratching moment for me as we stood there on the street about to part company once again, realizing that here we were all these decades later, our childhoods and the better part of our adult lives behind us, and there was not a single word left unsaid. It was simple, really: I had set out in my youth in search of meaning, and he in search of money, *and never the twain should meet.*

This Kipling-like phrase that describes the gulf between money and meaning really set the tone that next morning at the Slow Money National Gathering 2011 down at Fort Mason. Fort Mason, located about a mile north of Fisherman's Wharf, is a large military installation built a hundred years ago. That brawny, fin de siècle architecture, all fifty buildings, was impressively arrayed on a pine-filled promontory of land across from Alcatraz Island. There's an upper and a lower Fort Mason. Not having the information in front of me, I entered Upper Fort Mason and soon realized that I needed to get to Lower Fort Mason. Apparently at Upper Fort Mason military stuff still goes on, while down at Lower Fort Mason can be found all manner of nonprofit organizations and public programs.*

---

* Current organizations that can be found at Lower Fort Mason, according to Wikipedia, are BATS Improv, Blue Bear School of

Finally where I needed to be, I entered the main building and approached the reception table. I could see my name tag from three steps away. It was the only one with the words "One Day" written on it. The conference was a full three-day event, but I had made arrangements to attend just the first day, at a considerably reduced rate. I received my program and started to circulate around the room. There were long tables heaped with bowls of fruit and trays of bagels, Danish, and urns of coffee set up on both sides of the hangar-sized space. I steered clear of the food because I had already eaten a huge freebie breakfast back at the hostel. The main space was not too awfully crammed with people. There was a bookstore vendor right there just past the reception table, which drew me over. I perused the colorful jackets and catchy titles before drifting farther into the venue among the attendees.

I came to rest in a quiet spot over by an exit door with a view of the wharf. I stood there reading the program's introduction. The Slow Money National Gathering 2011 was described as offering its attendees "an opportunity to participate in an emerging national conversation about how we can fix our economy from the ground up as well as investment opportunities in dozens of enterprises that are rebuilding local food systems locally and across the country." I was very curious to hear what these

Music, SFMOMA Artists Gallery, Long Now Museum & Gallery, Magic Theatre, California Lawyers for the Arts, Greens Restaurant, Mexican Museum, City College of San Francisco Art Campus, Lily Cai Chinese Dance Company, On the Commons, Environmental Traveling Companions, Readers Book Store and Cafe, World Arts West, San Francisco Children's Art Center, Young Performers Theatre, and the Museo ItaloAmericano.

folks had to say, especially Wes Jackson, because I sure wanted a new economy to believe and invest in.

Of course I don't need to tell you, my joyful reader, that there is plenty of quantitative and qualitative research proving that money does not equal happiness. In fact, they've reduced it to down to an actual dollar amount beyond which point happiness is no longer affected—$75,000 per annum, to be exact.* And of course, it's not just money in the bank that correlates to happiness; it's also how much more wealthy you feel in comparison to the people around you.† I can validate these truths because I am an inordinately happy person, and between my wife and I, we make *about* that magic number; plus, the town we live in is chronically depressed, so relative to our neighbors, we're pretty flush. (Added to that is the fact that we live in America, with access to everything anytime we want it, most especially hot showers, orange juice, and music by Louis Armstrong.) That said, please don't think that I am not ever anxious about money, because that isn't true. I am very anxious about it in two ways: 1) Retirement. There's no way I have enough money now, nor am I on track to have enough money saved up in the future to meet the projected costs associated with old age and retirement. 2) Where the hell is my money anyway, and what's being done with it?

---

* Read more: http://www.time.com/time/magazine/article/0,9171,2019628,00.html#ixzz1immlr0un.

† Read more: http://www.time.com/time/health/article/0,8599,1974718,00.html.

Regarding retirement: From all the reading I've done, it seems pretty clear that my wife and I will need to have saved a minimum of $500,000 to be relatively "safe" in retirement. My round little nest egg at present is nowhere near that half-million mark, and I've only got fifteen years left until retirement. I certainly don't want to be a burden to my children, whose gainful prospects, by the way, are far worse than mine ever were. And though I am assiduously saving three dollars a day (that's eighty dollars a month), it still won't get me where I need to be fifteen years down the road or even fifty years down the road. Even more pathetic than that are the recent studies of wealth in America that put me, with my impecunious net worth and anemic salary, not in the lower class or the middle class, but in the upper twentieth percentile!* If the future is anything like they say it will be, many of us are going to be up the proverbial creek without a paddle, or even worse, gathering in the streets along with the other 99 percent in search of corporate CEOs, hedge fund managers, Wall Street traders, energy company executives, or investment bankers—their fresh heads to chop off.

So that brings me to my second money anxiety: If you've read any of the books by Michael Lewis, specifically *Liar's Poker* or *Boomerang*, then you'll know why I am stressed about where my savings dollars and retirement account money is going (puny as it is) and how it is being used.† Another great book that sets

* Check this article out: http://allfinancialmatters.com/2010/03/29/77-of-american-workers-have-less-than-100000-in-savings-investments/.

† If you don't feel like reading those books, which is understandable, as the mumbo-jumbo of avaricious hedge fund code talkers does not make for very interesting or uplifting prose, then peruse his essay "The End," found here:

the record straight about what's going on down on Wall Street with my money (and yours) is *13 Bankers* by Simon Johnson and James Kwak. In a nutshell, here's what they have to say: "The last few years have proven that our financial sector and its political influence are a serious risk to our economic well-being, and without significant change there is no reason to believe we will not soon experience the next boom, the next bust, and the next president explaining to the American people that he must rescue Wall Street in order to save Main Street."* If I am to believe what I read, then I can assume that all sorts of nefarious games are being played with my innocent, hard-earned savings. It's a despicable situation, and, like GMO crops, one of the major issues of our time.†

Anyway, these are the reasons I had come to the Slow Money conference. I was hoping to learn about a new way of investing that would keep my money out of the wrong people's hands and put it into the right people's hands, preferably small farmers'.

I walked along the wharf for a while, and by mere coincidence came upon Wes Jackson being photographed by a group of media people, with the Golden Gate Bridge in the background. I stopped about ten feet away and watched, but quickly felt uncomfortable, so I walked a few paces away and turned my back

---

http://www.greatertalent.com/speaker-news/michael-lewis-writes-about-the-end-of-wall-street/.

* Simon Johnson and James Kwak, *13 Bankers: The Wall Street Takeover and the Next Financial Meltdown* (New York: Random House, 2010), p. 190.

† By the way, my hard-boiled nest egg, even with the eighty bucks a month I've been adding to it, has remained virtually the same over the past ten years.

to them, looking out into the bay. I should point out that I actually had an introduction to Wes Jackson set up beforehand. I am a friend of his daughter, Sara. We met six years ago down in Louisville, where every June we, along with a thousand other English teachers from across the nation, subject ourselves to cruel and unusual punishment by grading the Advanced Placement English literature exam. I had known her for five years and had had no idea she was Wes Jackson's daughter, until the summer of 2011, when my travelogue *Seeds* came out. She wanted to buy a signed copy of it from me so she could gift it to her dad's friend who had invited her over for dinner in nearby Henry County, Kentucky.

"So I take it your father's friend is a real tree lover?" I asked her.

"I assume so. Of course he's an author, too; I'm sure you've heard of him."

"Who is he?"

"Wendell Berry."

I about crapped my pants. "You're giving my book to Wendell Berry?"

"Yup."

"How does your dad know Wendell Berry?"

"They go way back. My dad's a writer, too, but he's better known as an experimental farmer."

"Really? What's your dad's name?"

"Wes Jackson."

I about wet myself.

So these thoughts were flashing through my head as I stood watching Sara's dad being photographed at the end of the pier.

But then something caught my eye out in the water. You're not going to believe me when I tell you this, but it was a diving, flightless bird: a penguin, I think. I totally forgot about Wes Jackson and became obsessed with this thing. It would pop up for a moment near the quay, and then dive down again for minutes on end. As a bird-watcher, I know my birds, and this was no bird I'd ever seen outside of a zoo. It was a weird-looking thing. It had feathers, but its wings were stubby and undeveloped, like turkey wings fresh out of the oven. I thought it was some kind of baby bird at first, but the way it dove so skillfully . . . no, this was no baby.

Every person I ran into I asked about it: "Say, do you happen to know if they have any reports of escaped penguins from the zoo living wild in San Francisco Bay, because there's one right out there now. See it?"

Before I got in too much trouble, the program started, and I had to hurry back inside to hear the speakers.

Woody Tasch, founder and chairman of the Slow Money movement, spoke first. He talked at great length about the growth of the movement over the past three years (the first national gathering was in 2009 in Vermont) and how millions of dollars had been invested in small food enterprises around the country since then, and how new chapters of Slow Money investment clubs and an entity called the Soil Trust were opening up nearly every month. "Slow Money was evolving quickly from a movement to a revolution," he proclaimed.

At this point, instead of trying to re-create his speech of that morning, I have opted instead to include here a much snappier summary of both Woody and the movement, lifted from the prologue of his book, *Slow Money*:

> The problems we face with respect to soil fertility, biodiversity, food quality, and local economies are not primarily problems of technology. They are problems of finance. In a financial system organized to optimize the efficient use of capital, we should not be surprised to end up with cheapened food, millions of acres of GMO corn, billions of food miles, dying Main Streets, kids who think food comes from supermarkets, and obesity epidemics side by side with persistent hunger.
>
> Speed is a big part of the problem. We are extracting generations' worth of vitality from our land and our communities. We are acting as if the biological and the agrarian can be indefinitely subjugated to the technological and the industrial without significant consequence. We are, as the colloquial saying puts it, beginning to believe our own bullshit.*

Woody was a percolating, gyrating bundle of energy. He moved constantly up there at the podium, rolling shoulders, swaying from one foot to the other, scratching at his big curly tuft of hair. Then he introduced Wes Jackson. There was a great

---

* Woody Tasch, *Inquiries into the Nature of Slow Money: Investing as if Food, Farms, and Fertility Mattered* (White River Junction, VT: Chelsea Green, 2008), p. xvii.

round of applause among the shiny, happy people now jam-packed into the hall. I stood up, moved closer to the stage, and got my camera out.

Like all of the farmers I had visited, Wes Jackson is the salt of the earth, and as true an American treasure as we've got. Standing up there with his big-toothed smile, full head of gray hair, and sweet down-home accent, he minced no words about our present predicament (on a global basis, soil erosion is accelerating), who's responsible (we all are), and what needs to be done to fix it (supplant annual root crops with perennial root crops). He talked a little about his life, his football-playing days, his home and family in Kansas, and what his famous Land Institute is doing with its perennial wheat project. He told about how he and Wendell Berry, and Fred Kirschenmann from the Leopold Center for Sustainable Agriculture, had recently gone down to Washington, D.C., and presented the secretary of agriculture with their own version of a fifty-year farm bill. It made my heart sing to think of these wise men marching down to Washington to teach our government how to save the planet's long-term food supply. In the end, he summed it up the same as he did in his most recent book, *Consulting the Genius of the Place*, saying: "It is redemption time, and we have to redeem ourselves as we redeem agriculture. By starting where our split with nature began, we can build an agriculture more like the ecosystems that shaped us, thereby preserving ecological capital, the stuff of which we are made, and guaranteeing ourselves food for the journey ahead."*

---

* Wes Jackson, *Consulting the Genius of the Place: An Ecological Approach to a New Agriculture* (Berkeley, CA: Counterpoint, 2011).

There was a huge standing ovation, and the keynote address was over.

A few minutes later, they set him up at a book signing station outside the main stage area, and I waited at the back of the line with my copy of *Consulting the Genius of the Place* for him to sign. When I finally got up to meet him, face-to-face, I forgot what I wanted to ask him, so I just said: "So you're Sara's dad?"

"That's right. You know Sara, do you?"

"We correct the AP exam together down in Louisville."

"That's some tough work you guys do down there."

"You got that right." I nodded.

He was just about to sign the book I had handed him when he suddenly looked up at me with a tilted head and said, "Say, aren't you the fella who wrote that book *Seeds* that she gave me for my birthday?"

"She gave you *my book* for your birthday?"

"Yup." That was his response. I wasn't sure whether it was a good *yup* or a bad *yup* or an indifferent *yup.*

I really wanted to say something important to him at that point, but absolutely nothing came to my mind, so I went with the first thing that did: "I know you used to live in this area and teach here a while back, but do you happen to know anything about the kinds of birds that live in San Francisco Bay because I saw this flightless thing swimming and diving out in the harbor that—"

"No, no, I don't know anything about that. Nice meeting you." He cleared his throat and looked at me askance while also indicating that there were people behind me who were waiting to get their books signed.

I slinked off into the crowd.

There was a bagged lunch provided, and I ate it outside in the sun, *o sole mio*. I did a pretty good job, too, with my back to the water, pretending that I was not looking for the flightless bird that kept bobbing up and down in the margins of my mind. Of course, it was a great sandwich provided by local farms, all fresh and organic, with plenty of exotic herbs and spices, fruits and nuts. And I only asked maybe three or four perhaps half a dozen people if they knew anything about the penguins living in San Francisco Bay. No one did.

After lunch there was a breakout session that I wanted to attend titled "Financial Planning and Slow Money: Where does Slow Money fit in your portfolio? Four investment managers give their perspectives on Slow Money." I sat in the back and listened. At first it sounded to me like Slow Money was more a charity than anything else. But as the conversation evolved, the one question I was dying to ask finally popped out:

"I have a retirement account, and I was wondering if Slow Money offers retirement account services, with IRAs and 401 plans? I'd rather have my money in Slow Money than in who knows where."

The four speakers seemed nonplussed by the question, but then, like Superman, the woman seated in front of me stood up and proclaimed that indeed there were retirement programs available so that people could invest their retirement portfolios in small farms. This was Leslie E. Christian, the CEO of Portfolio 21 Investments. She is a well-known investment professional committed to building the investment case for local economies with a view on ecological limitations. I took her name and number and headed out to the car.

Without even a last look out toward the harbor, nor even at Alcatraz scintillating between San Francisco and me, I headed northeast, toward the walnuts.*

* If, by chance, my magnanimous reader, you are an ornithologist who specializes in the flightless birds of the Bay Area, could you please contact me at the following address: rhoranwrites@yahoo.com? It would sure settle a lot of commotion in my life. Thank you.

Walnuts

SIERRA ORCHARDS (ORGANIC)

WINTERS, CALIFORNIA

OCTOBER 2011

There's a famous commencement address attributed to Kurt Vonnegut that he didn't write or deliver. Bogus though it may be, there was a line in it that always stuck in my craw: "Live in Northern California once, but leave before it makes you soft." I'd spent a couple of summers in Southern California during my late teens and early twenties, and I could understand how that axiom might apply to living in that area of the country, what with the sun and beaches and surfing and hot babes in bikinis and driving around in convertibles. But why Northern California? And as I drove north on the Oakland side of the Bay Bridge toward Winters, it became clear: *because Northern California is practically perfect in every way.*

Except for the heavy traffic between Berkeley and Vallejo, the world outside my car was an explosion of beautiful pastel colors, mixed with earthy tones and natural fecundity. Walnut groves marched off toward the distant mountains like a parade of wooden soldiers. And what mountains! Beyond Vacaville, the skyline was full of stately dark masses looming high up ahead like giant theatrical stages for the gods. And the interstate seemed to be leading me and all the other demigods directly to them.

Then I was there. A mere mile from the exit, I saw the number on the mailbox. I hadn't thought about what to expect. To tell you the truth, I didn't even remember how, exactly, I had connected with the owner of the Sierra Orchards. Initially I had wanted to harvest peanuts down in Alabama, but none of the good old boys who farm the goobers down in the Yellowhammer State would have me—no way, no how. But I had to harvest nuts; that was a must. I was coming to Northern California to harvest grapes anyway, so somewhere along the line the walnut harvest fell into my lap.

I drove slowly into the driveway, my mouth opening wider as I advanced. To the left was a yummy manor with a tall roofline, cradled within a bright copse of exotic trees, like pistils and stamens in the center of a rose. There was a pool out in the back, bright flower beds every which way. There was nothing loose, nothing out of place. This was truly the Bizarro World version of the Wood Prairie potato farm. The long driveway wound around the back of the place, resolving into separate driveways that led in several directions. I slowed, trying to decide which way to go. It didn't really matter because they all eventually led into the walnut grove. It was an odd sensation being trunk-level in a grove that large. Looking into the ceaseless understory of the trees, it felt a bit like I was standing in a room full of mirrors, seeing one image repeated infinitely in every direction. The boles of the trees were odd-looking. Their bottoms were dark, like they were wearing black pads similar to the ones on goalposts. The first three feet of each trunk was black, and then gray above. I soon found out that these nonnative walnut trees had all been grafted onto the trunks of native walnut trees.

Suddenly a man on an ATV with a blue tailored shirt came racing around the corner of a large warehouse to my left. I saw him for just an instant before he made a hard U-turn and disappeared back behind the structure. I pulled the car over to the

edge of the driveway, near a gate that led to the house, and parked. I glanced at my watch. It had taken me just an hour and fifteen minutes to drive from the Bay Bridge. An instant later, the same man came racing around the same corner on his ATV. This time he pulled up and leapt off the vehicle, then bounded straight up to me, followed by a curly, white-haired little mutt that must have been riding on the seat next to him. Craig McNamara was about my height, five ten, late fifties, grayish-white hair, wire-rimmed glasses, well-balanced physique, sky-blue eyes, and a smile that matched the Northern California weather. I had to fight the urge to look over my shoulder to see if this shining, congenial welcome I was receiving wasn't meant for someone behind me. Nope—all mine, and its ardency never flagged the entire time.

We shook hands.

"This is Teddy," he said, referring to the uncut poodle sitting smartly on the black tar next to his feet. "He's the foreman around here, and way too clever for his own good. So how was the conference?"

"Wonderful. Very informative. I got to meet Wes Jackson."

"Wes Jackson! Wow! Wow! How great is that!"

He stood there beaming at me. I have to say that it was the perfect thing to say because I was feeling a bit contrite about changing the date on him so last-minute. (Originally, I was scheduled to arrive the day before, but when I found out about the Slow Money conference, I contacted him to tell him I'd be there a day later.) He put me right at ease, mentioning the conference and making a big deal of it. I would learn that that sort of thing was Craig's gift—the ability to make all in his presence feel comfortable and important.

"Well, let me show you around. Hop on."

And off we zoomed, I on the back of the ATV, my legs dangling over the sides like I was sitting on the dock, and Teddy tightly seated next to me.

We whizzed around the large warehouse and came to rest in front of three tall garage bays, with a two-story-tall sorting machine in the far left one. It was quite a contraption. At the far end of the machine, where the walnuts came out on a conveyor belt, four men stood on either side of a picking station, sorting out walnuts. One man was up on the second level of the machine, which reminded me of a much heavier version of the Great Gene Cranberry Sux Machine, and another man was in the middle bay, raking a large pile of walnuts down into a grate-covered hopper on the floor. It was deafeningly loud inside. All the men wore white jumpsuits, gloves, and earplugs. All but one were Latino.

Another man appeared from behind us and came over to Craig. He clapped him on the back, said a few words in Spanish, and then introduced us. He explained a little bit about what I was up to, gave a nod of his head, and then led me around to the back side of the warehouse.

"That's Rodrigo. He's the foreman. I have four full-time people and twenty seasonal workers."

He climbed up a flight of stairs and was out of sight. I chased after him and found myself up under the roof of the warehouse, walking along a catwalk and looking into six huge holding bins, six feet long and who knows how deep, swelling with walnuts. Craig had walked the length of the catwalk and disappeared before I was even halfway across. He was in constant motion, that man, and like all the farmers, there was never a wasted moment

or movement. I took a deep breath, lengthened my stride, and bounded after him.

When I climbed down he stood there waiting to explain it all.

"We don't have storage here. We have to keep things moving. The walnuts come in from harvest at thirty-five percent moisture, and we dry them down with propane to eight percent at a hundred and ten degrees."

He walked me over to two huge trailers parked side by side. They were loaded with nuts, drying. He walked over to the ATV and removed what looked like a laptop computer, then came around to the side of the closest trailer and plugged it into an outlet on the side of the metal tank.

"Conductivity allows the measurement to take a moisture readout. These are almost ready. Hmmm. This is where I earn my keep. The moment of indecision, whether to take a load over to get processed now or to hang on to it one more day until it is sufficiently dry. Because if it's not, then you run the risk of having it rejected."

"If it's not dry enough, they reject it?"

"Heck yeah! What the processor does is take a hundred nuts as a measurement to grade them, showing the full spectrum of good to bad by a third-party evaluator. I'll show you."

He marched me over to his office, which looked like a little bathhouse. It was fifty paces from the warehouse. It had big bay windows on three sides, all of which looked out into the enchanted little grove. He had a huge desk that covered one side of the room, neat but loaded with papers and books and stuff. And next to that, a beehive under Plexiglas. The bees were pretty damn busy in there. He saw me looking at it and came over and stood by my side, pointing.

"See that one?"

"Yeah."

"That's the queen."

"Interesting. So are the trees pollinated by bees?"

"No, it's wind pollinated. I just enjoy having them here in my office."

I stood there with my arms dangling self-consciously at my side as he went back over to his desk and rummaged through papers. Suddenly, and with a flourish, he yanked on a sheet of paper and held it up in front of me. It was a computerized grade sheet that read:

VARIETY: HARTLEY

EXTERNAL		KERNEL	SERIOUS DEFECT	
Jumbo	50		Insect	1
Large	30	Color–light	Rancid	0
Medium	13		Mold	8
Baby	7		Shrivel	0
Peewee	0			

"It's all about edible yield content versus average defect. Pests are the biggest problem. I hire a pest control advisor. This is one reason why we're organic. Economically it makes sense because the organic market doesn't pay on size like the conventional market does. But more important, I'm organic because my kids and wife live where we farm."

I stopped reading and looked out at the grove.

"So you noticed the boles of the trees?" he interpreted.

"Yes."

"Terrific. Most of these trees are fifty years old. They're from China and India and France, but they were first grafted onto native black walnut trees. Grafting is a real art form, and of course it's the only way these groves would exist, because grafted walnut trees are hardier and produce more reliably than trees planted as seedlings. Plus, they produce earlier crops of higher-quality nuts."*

"I thought I saw some regular black walnut trees when I was coming in off the interstate," I responded.

"You saw those? Excellent. Good eye. There are sixty-seven different varieties of walnuts in the world. The original California growers were Spanish missionaries who grew English walnuts or Persian walnuts [*Juglans regia*] along the California coast. We grow four varieties here: Hartley, Franquette, Chandler, and Howard. Franquettes tend to be darker than the rest in color, with a meatier flavor. I don't know if you could tell the difference up there, but I'll show you later."

He had a teacherly manner, and he was especially good at complimenting your comments and questions, encouraging more out of you.

"How many acres do you have?"

"Four hundred and fifty, but it's not all walnuts. I'll show you what else we've got."

He marched back outside and I followed. We hopped back on the ATV and sailed off into the grove. He took me to two

---

* Walnuts contain some of the highest amounts of antioxidants of any fruit, nut, or vegetable, which lower cholesterol and fight heart disease. The leaves have been used to cure herpes and ringworm, and the hulls have been used to treat liver problems and to help people lose weight.

different areas: the new and the old. In the older area, the trees were more spread out, at 55 per acre, whereas in the new area they came in at 187 per acre. A different look and feel altogether.

Then we went back to the warehouse and got into a truck parked next to a huge bank of solar panels, which were angled up toward the sun in an attitude of technological sunbathing. I counted 111 three-by-five-foot panels. It was quite an impressive sight.

"What about those?" I motioned.

"Oh yes. They're working like a charm. The solar panels produce twenty-four kilowatts of electricity and power our hulling operation, well, and home. I'm very pleased that we installed them and definitely would like to put in another solar-panel system to operate our hundred-horsepower pump that irrigates two hundred and fifty acres of organic walnuts."

We loaded into the truck and he drove us out of the driveway and down the tree-lined road. All those neat rows of trees, and the sun, the soft colors, it was such a stark contrast from out east. I couldn't help but think about Jim Gerritsen and the Wood Prairie Farm and all those rocks and dirt and damp and wilderness and moose . . .

The road we drove along, in through the trees, was narrow but perfectly straight and smooth as could be, which is probably why we kept passing packs of cyclists all decked out in bright gear and helmets and eyewear, heads down and really pedaling. It was perhaps a mile down the road when we turned off into another walnut grove, with a large, square hacienda in the middle. The house was older than his yummy house, but all renovated and freshly painted.

"My son Sean lives here, but I don't see his car. We'll see him tonight at dinner." He drove around the back. "This is the Center for Land-Based Learning."

"The what?"

"You don't know about this?"

"No."

"It's a nonprofit educational program for high school students. We work with thirty public schools in the area to provide training for future farmers."

"Wow, what a terrific program. What sort of stuff do you do?"

"This is a six-acre CSA farm run by a young farmer who's doing a great job bringing back biodiversity and beneficial soil management. It's a nice corridor, too, between the walnuts and fruit trees. And this is what we call the Beginning Farmer Training Center. The objective really is to help overcome barriers, especially in the underrepresented populations, and to build community."

He parked the car and we got out. I was pretty overwhelmed at this point. Too much information coming at me all at once. There were buildings out behind the main house that looked like outdoor classrooms. There was even an outdoor kitchen and dining area, something completely off the charts for an East Coaster like me. The six-acre field to the west of us was plowed, but there were no crops in it. A man about Craig's age was in the outdoor kitchen. Dressed in overalls with an extremely sinewy build, he was installing a dishwasher *inside the outdoor kitchen* (also completely unimaginable out east). Craig introduced us. The man had a very thick French accent. I was going to say something to him in French, but thought better of it. I didn't want to have to explain myself in front of Craig. (Because I used to live in the French-speaking part of Belgium, I speak French. However, I have been told by French speakers that when I speak French, I sound like a Frenchman, but with a slight Argentine accent. Probably because I speak Italian. On top of the fact that I'm American, it has caused me no end of trouble whenever I meet and speak French to a native Frenchman.)

"The Center for Land-Based Learning is part of the California SLEWS program, which stands for Student and Landowner Education and Watershed Stewardship. We have six locations throughout the state, but this is the home office, here at Putah Creek. There are twelve full-time staff members; not all are certified teachers. But they're teachers. They're life teachers."

"Putah Creek. That's quite the name."

"It's actually a native word, and not the one you're thinking."

"Oh. So what about you? How do you fit into the organization?"

"I'm the president and founder," he said *avec nonchalance*.

The CSA field was the only treeless expanse of land that I had seen since arriving. Beyond and all around were the ubiquitous walnut trees, all about the same size and color. Again, this Northern California state of mind puts you in a trance where you begin to expect all things to be neat, orderly, and glistening under the ripe red sun. I had experienced an orderly forest once before, back in Belgium. The once wild region of the country called the Ardennes, where the Battle of the Bulge was fought, is today a virtual nursery of conifers, all planted equidistantly apart, that rolls for miles and miles across the countryside. For a person who has hiked the wildest section of Appalachians through northern Maine, it was the strangest place I'd ever experienced—a manufactured forest!

Next thing I know, I'm back at the main house, being shown "my room." Formerly the bedroom of the McNamaras' older son, it was nothing big or opulent, but what it lacked in size and style it more than made up for in historic documents. I kid you not, my loyal reader, above the bed, neatly framed and mounted, were two side-by-side photographs of the late Robert McNamara, Craig's father, at the White House walking with JFK in the left-hand photograph and LBJ in the right-hand one. And under each photograph there was a scrawled (you couldn't possibly read what it said) message and a signature from the two presidents. I knew about Craig's father, having read about him on the Internet, but Vietnam and JFK and LBJ and McNamara himself were before my time, so it didn't really make much of an impact on me, until the photographs . . . Hanging there in crisp, clean wall mountings above the bed I was going to sleep in, the image of his father with the two former presidents transformed the abstract Internet version of history into real life, and chastened my proud heart.

I had my own bathroom! There was a large tiled shower, towers of towels, and plenty of interesting reading within reach. I was told we were going out to dinner, so I quickly showered and got ready. Under the hot water, giddy with good luck, I considered my silver-lined fate as nothing more than an accumulation of karma. However, finding myself at this particular farm, with this particular family, I also understood that it would not be all showers and dinners out. Being in the company of someone like Craig McNamara, who was not a simple farmer, I would have to do more than participate and observe—I would have to understand. That was the key to working with people at the top of the food chain: they don't have time to waste on fantasy-chasing, slow-to-catch-on, ragged individuals like me, which is the main reason why people like me end up "poor, ostracized and disgraced."* When I came out into the kitchen, Craig was standing there at the counter, reading my book *Seeds*, which I had gifted him.

"Wonderful stuff. Wonderful. All these historic trees that the people had connections to . . . what a great idea!" He began telling me about the trees in the Arlington National Cemetery in Washington, D.C. I didn't have the heart to tell him that I had no cemetery trees in the book simply because it was about the living connection between people and trees, and not in death. When he finished the story, he clapped his hands together and said, "Come on, let's go eat."

The town of Winters looks like a Main Street movie set—the twinkling streetlamps, the inviting storefront windows, the cluttered sidewalk displays. We parked right at the corner of the sleepy four-way intersection, as much a crosswalk for pedestrians as a thoroughfare for cars. He pointed at the place we were going into—very swank—and immediately popped out of the car and strode down the sidewalk and around the corner. I jogged after him, and didn't catch up to him until he was sliding into a luxurious booth at the back of the restaurant. A waitress stood there waiting for me.

---

* The full quote is from Lewis Lapham, who once wrote: "Except in a few well-publicized instances (enough to lend credence to the iconography painted on the walls of the media), the rigorous practice of rugged individualism usually leads to poverty, ostracism and disgrace. The rugged individualist is too often mistaken for the misfit, the maverick, the spoilsport, the sore thumb."

I slid into the seat across from Craig, opened the menu, and glanced at the prices. *Ouch!* I felt that sweaty, itchy feeling ripple across my collar. Craig didn't even bother to pick up his menu. When the young waitress was gone, he glanced over at me, drumming his fingers on the thick tabletop:

"So, do you like lamb?"

"It's my favorite meat."

"Then put the menu down: you have to get the rack of lamb here. It's the best, I'm telling you, the best!"

"That was easy." And I meant it. Amazingly, the itch in my collar had vanished completely.

Just about the time the food arrived, Craig's son Sean appeared on the scene. In the full vigor of youth, he had an athlete's build—like a cornerback on the football team or a middleweight wrestler. He had a thick neck, a thick chest, and a thick head of curly black hair. But his most striking feature was his huge brown eyes. I hadn't seen peepers like those since I stopped watching *Speed Racer.* Two minutes later Craig's beautiful wife, Julie, fresh from her ukulele club meeting, arrived. Julie struck me as someone with not only the gift of grace, but also an actress's timing and savoir faire. She wore black-rimmed glasses, and had blond hair and smooth attractive features. She reminded me of a cross between Meryl Streep and Shirley Jones.

For several minutes it was as if a carnival had descended on our table. All this wild, creative energy was dancing and whirling around, beating drums and blowing whistles and shaking rattles, fueled by love and companionship and mutual admiration. It was a family affair for sure, and a familiar scene for me, because at my house, when all of the losers get together, there is this same carnival of mutual delight. I sat there contentedly, drinking my beer, eating my lamb, and feeling like one of the family. But then the wild chatter and laughter and joking suddenly died down, and their huge McNamara eyes were on me. I sort of half-expected it. They were too bright, too intelligent, too aware of themselves not to possess that dreaded quality that all writers fear in their subjects—self-effacement.

It was a bit awkward at first, to be the cynosure in the constellation of this illustrious family, but I acquitted myself fairly well. What helped was having a huge plate of lamb chops in front of me, so that when I was at a loss for words, which I am chronically, I would take another bite and deliberate over a long, eyebrow-scrunched chew. I was fielding a lot of questions, mostly about my harvesting adventures, but also about my writing, my family, and my years living outside the United States. They weren't nosy or prying questions, of course; they were polite questions perfectly worded to make me feel welcome and important.

But when I got to the part about hiking the Appalachian Trail and traveling around the country for a year after college, Julie insisted that Craig swallow his pride and tell me *his* after-college travel story. It was the most outlandish personal adventure I'd ever heard, and I've heard a lot. It must be mentioned that this story he told took place during the peak of the Vietnam War, when Craig was in his early twenties. It should also be noted that the Vietnam War, a war Craig vehemently opposed, was also called McNamara's War. The story went something like this:

When Craig dropped out of the University of California–Berkeley after two years, suffering from debilitating stomach ulcers, he and two friends bought motorcycles and rode them all

the way to Colombia. I forget how many months he said it took them. Once in Colombia, the friends decided to start a business, but Craig continued on alone. Leaving the friends and motorcycle behind, he traveled by foot through Ecuador, Peru, and Bolivia all the way to the southern tip of South America, at Tierra del Fuego. Back in Santiago, Chile, he met an Easter Islander who invited him to live with him in his country. So he went. Easter Island is famous for its huge stone statues that gave Thor Heyerdahl the idea that the South Sea Islands hadn't been discovered by people sailing from Asia, but by people sailing from South America. And he lived there among the natives for a year, like Herman Melville in *Typee*, starting his own dairy cooperative as he slowly found his calling in life. On Easter Island his adoptive mother, Mama Vera Hito, gave him the name *Tukoihu. Tukoihu*, as legend has it, was an *Ariki,* or high chief, which actually made the *Maoi* move or walk from the volcanic stone quarries where they were carved to their *ahu,* or altar sites. This was an amazing feat because *Maoi* statues weigh as much as eighty-five tons and stand thirty-three feet high. After two years in a world far, far away, he returned home to California. He soon enrolled at the University of California–Davis, got a degree in agriculture, eventually bought the walnut farm, married, and raised a healthy, happy family, and was alive and well enough to tell me the story.

There is nothing so revealing as going out to dinner with people, especially people you don't know, and seeing how they react when the check comes around. It has been my experience that a real scrooge will pretend the bill is not even there, and when you reach over to take responsibility for it, he will feign surprise, but brook no interference, before relaxing back in his seat all comfort and smiles. Most people, however, no matter the circumstances, will scratch or rub their noses; make strange faces or odd sounds. And when you reach for the check, they will stumble for a response as if they'd never had to consider something so odd as this before. Some will suggest "helping out," others might insist on splitting it down the middle, and still others may want to add it all up and count it out to see exactly who owes what. It's the rare person who growls as you reach for it, like a dog when a hand threatens to take away its bone. *"Don't touch it!"* he menaces. Needless to say, Craig represented the latter.

The next morning I wandered out to the light and airy kitchen and found a skillet on the professional-grade stove, a spatula nearby, a plate, a cup, a dozen fresh eggs, an urn of fresh coffee percolating, English muffins, assorted jams, butter, and utter silence. I waited at the kitchen table, drumming my fingers, for about ten minutes before remembering Craig's last words before I went to bed: "Consider yourself at home." That was enough to make me reach for the spatula. Ten minutes later, as I was making my way through two fried eggs, an English muffin, and a large cup of coffee, Craig bounded into the kitchen from outside full of that endorphin-charged early morning soar. He stood for a moment, considering the scene. His expression changed from surprise to admiration in less than a second.

"You helped yourself? That's excellent. It's what I keep telling people around here; you've got to take initiative."

Then it was his turn to make himself some breakfast. I sat and watched him do it. Again, not a wasted movement. He was like a gymnast on the parallel bars, all clean lines and level

swinging motions, and no loose knees or awkward shifts. He brought his food over to the table and we sat together and chatted. He told me that Julie was flying to Rhode Island that afternoon to spend time with their daughter, who was at college there. He had to drive her to the airport, and after that go to some function at UC–Davis. In a nutshell, I would be on my own for dinner and afterward. As he was telling me this, I spied a list in his hand, that irreplaceable emblem of well-managed farms. I enquired about it, and he shared it with me. First thing on the list was hauling a tanker of walnuts over to the processor.

Ten minutes later I was seated on the tractor above and behind him as we rolled down the road toward the processing plant five miles away, towing one of those huge tankers of walnuts behind us. On the ride we discovered we had a mutual acquaintance. Hanging on the wall opposite the photos of LBJ and JFK in the bedroom where I slept, there was an event poster from Martha's Vineyard. It was an area of the island that I'd been to many times during college. I had a very rich friend whose parents owned a piece of property on a private beach there—an amazing place. I told him about it. He shot me a funny look and then asked the name of the family.

I was about to tell him when, suddenly, Sean drove by us in his pickup truck, waving wildly and yelling out the window, the pickup swerving from side to side.

"Crazy man!" Craig shouted at him; then he turned up to me. "He was out late last night. But tell me, what was the name of the family?"

I told him. He let his foot off the gas: "That was our property. My parents sold that land to them," he said in disbelief.

"It's a small world," I responded.

Suddenly there was a great shout above and behind us. I turned around and was dumbfounded to see Sean sitting on top of the walnut tanker, looking down at us with a Cheshire Cat grin.

"Crazy nut! How did he do that?" Craig howled.

"I have no idea!" I shouted over the roar of the engine.

Sean must run like a deer to have parked the truck, jumped out, raced after the tractor, caught it, and climbed aboard in under sixty seconds. Amazing. Sean was turning out to be a kind of trickster figure who would appear all of a sudden, larger than life, then be gone.

The processing facility was quite an interesting place. The amount of complex machinery and the size of the warehouses was something to see. They had a gasifier or cogeneration furnace that ran on biomass, that is, the shells and skins of the walnuts it processed. The thing looked like something NASA might have built. We pulled right onto the scale. They weighed everything, and within seconds it was all being unloaded and processed into bins. Sean, more than Craig, explained a lot about how the processing system worked, how they took the hundred sample walnuts and ran them through the machine's stainless steel torso with its glowing red and green digital numbers. Craig at one point mentioned a concept that I'd never even thought about: "Color is important in agriculture. Most often light is right." Then he commented on the walnut varieties: "Hartley's have 42 percent meat to shell. The Chandlers are 55 percent meat to shell." He also told me that the big worry is anaphylactic toxins, which cause anaphylactic shock and can infect walnuts. So can E. coli.

I expressed the thought that it was very convenient for them

to have the processing plant just down the road from their orchard. But he confessed that the price he got from the owner was a constant source of anxiety because the guy didn't pay as well as other facilities throughout the state, but then again he was right next door. And the worst part was, Craig had to sign a contract a year prior to using the facility. There is no paradise.

While Craig drove the tractor back alone, Sean and I ended up getting a ride to his truck from the foreman of the place, who hailed from my neck of the woods, central New York and the land of Oz.* Interesting guy, a sort of salty-dog sailor type, but a landlocked version. And on the ride back to the house, Sean and I got to know each other a little better. He had gone to school in Washington, D.C., and ended up in Baltimore working as a carpenter for an architect who "built in the vernacular." I had to think about that expression for a minute before it clicked, but coming from a twenty-four-year-old dressed in sandals, shorts, and a T-shirt, it sort of took me by surprise. He loved Baltimore, loved the urban decay, the hardscrabble, bare-knuckle quality of life there. He also proved to himself as well as to others that, contrary to popular belief, not all people from Northern California are soft, and that he could move and shake in a tough world. But something called him back home, back to the orchard and to the walnuts and to the trees along Putah Creek. I understood this implicitly to mean that the pull of the land and the soil and the crops is so great that no matter how gifted or how full of wanderlust or even how youthful you are, you are drawn back to the land as if it were full of sweet-singing Sirens. We

* Oswego, New York.

agreed to meet that night at the music hall in Winters. And then, like the trickster, he was gone again.

Craig arrived back on the tractor soon after and put me to work. It was one of those swollen moments that will be freeze-framed in my mind forever. He drove me on the back of the ATV, with Teddy sitting next to me, to an area in the middle of the orchard. He let me off, handed me a rake, and told me to find the fallen walnuts and rake them into straight lines like they were one row over, and to remove the large twigs as well. He said the crew was somewhere out there, and sooner or later they'd find me, pointing to the straight line of walnuts one row over as proof.

"This is the Zen part of walnut harvest, but it's also a very important part of it. You'll see when the crew comes how it all fits together. I'll be back at lunchtime to get you." And off he and Teddy zoomed.

I stood there under the equidistant trees, looking off into infinity in all directions. The sun was a soft, magisterial presence above. Exotic birds chirped and occasionally flitted in and out of my ken. The air was sweet and the temperature a soothing and caressing 80 degrees. The ground was lush and ripe and jumbled with fallen walnuts. I began to rake them into a long, thin line, slowly but steadily. In no time at all I had entered the Zen zone. I was ambling along the shore of my own personal Walden Pond, one with it all, a low bass line vibrating up in the limbs of the trees, a sweet flügelhorn singing the melody in front of fat, rhythmic chords. Suddenly I heard a noise. I looked this way and that—nothing. I returned to the Zen . . . then . . .

"WWWRRRRRRRR!"

I practically jumped out of my shoes! A UFO-like vehicle

appeared out of nowhere, seemingly hovering along down the row next to mine, extremely low to the ground, maybe three feet high, with a huge black circular brush seven feet long and two feet in diameter spinning and spitting walnuts and twigs out of its way in the lane in front of it. It was there for an instant and then it was gone and out of sight and sound. I stood for the longest time listening and not able to hear it off in the distance. What a machine! I walked over and examined the row that it had cleared. The brush in front was angled so that it swept all the twigs and walnuts to the right in a haphazard sort of line, hence the raking. But how did the walnuts and twigs end up on the ground, I wondered. It would take a few more hours to find out.

Not quite an hour later, I looked up and saw a body raking way up the way, a hundred yards or so. I wasn't sure who it was or if he was part of a crew or what. I forgot about him after a few seconds and went back to the Zen of walnuts. Then a few minutes later I heard a voice close by and looked up and was surprised again to find a crew of eight men with rakes not fifty feet away. There was something about the ground and the trees that must have vacuumed sound up and away so that there was no residue of it in any direction. They were all Latinos, and they barely paid me any notice.

"Hola, me llamo Ricardo. Soy un escritor. Escribo una historia acerca de sus nueces." (My name is Richard. I'm a writer. I am writing a story about your walnuts.)

They all started to laugh uproariously. One of the older guys said something to another guy and I understood that to mean that he wanted to know whose nuts I was writing about. There were great peals of laughter all around. Then another comment was made. I'm not sure, but I think the guy might have said something to the effect that I should probably write about Pedro's nuts *porque el es el mejor sujeto para la historia* (because he is the better subject for the story). There was uncontrollable laughter at that point, with some of the men bent halfway over and leaning on their rakes for support. You can imagine the look on my face. Oh yes, with my level-one Spanish, I was *carne fresca sin duda* (fresh meat for sure). In any case, I had made my presence known.

I worked right along with them. All the men were of Mexican lineage. I worked next to the oldest of the crew. I'm not sure, in retrospect, whether it was my own personal choice to work alongside that man or whether they sort of choreographed it; nevertheless, it was the exact right pairing. This gentleman was perhaps sixty-five years old, formal in his speech and manner, and not quite so bawdy as the others, who ranged in age from eighteen to fifty. While they were a very talkative bunch, boy, could they rake. I was amazed at how quickly they raked the walnuts into thin lines, and now that I think of it, probably the reason I was paired up with that oldest gentleman was a simple matter of speed—he and I being the slowest. At one point the man asked me the perfect question:

"Por qué tu estás trabajando y no sólo mirando?" (Why are you working and not just looking?)

I couldn't think of an answer for a while, but he waited patiently as I fumbled for words, first in English and then in Spanish.

"Porque yo puedo aprender más." (Because I can learn more.)

Soon another machine came along. Actually it was a series of three machines—a train of machines. First was the big green tractor. Behind it was a blue siphon/separator. And behind that,

like a caboose, was the sleek red trailer. The tractor pulled the siphon/separator as it sucked up the thin line of nuts, separated the twigs out, and vacuumed up the nuts into the trailer. Very similar to the potato harvester, it had a system of gates that culled the twigs from the nuts and accumulated them in a separate hopper on the side that could be emptied when full. I was dazzled by the series of machines, but not for long, because the man driving the tractor quickly eclipsed my prepossession with the machinery. He wore one of those white jumpsuits like the men at the walnut sorter, as well as a tall cowboy hat. I was struck by how regally he rode, tall in the saddle, if you will. And as he came abreast of the crew, they directed great caterwauling and jawing at him. It was not rancorous or critical, but high-spirited and full of fun and games.

And then I realized they were telling him about me, *el escritor, allí* (the writer over there) and *el está escribiendo una historia sobre tus nueces* (he's writing a story about your nuts). I was at the far end of the line, and he was searching the cast of dark-haired, dark-skinned characters, looking for an unfamiliar face, wondering what they were talking about. But then he spotted me, the very last one in the line, my head down, raking, trying to elude detection.

Pedro Garcia is that singular archetype found in all cultures and all literatures throughout the ages. He is the hysterical protagonist of the commedia dell'arte, the picaresque character of our fables represented by the likes of Don Quixote, Bugs Bunny, and Ed Norton, part genius, part buffoon, and part saint, around whom the human drama soars and swirls and hiccups. He pulled up alongside me on the tractor, turned it off, hopped down, waddled over, and accosted me. He was about sixty, five foot six, with blue eyes and reddish hair (very unusual for a Latino from Mexico), no front teeth, and a smile like a googling, milk-sated infant. His eyes twinkled. I kid you not, my bedazzled reader, I understood every word, every idiom and metaphor of it. I can't repeat it in Spanish, but in English it went something like this: "Okay, my friend, what's all this talk about your writing stories about my nuts? Well, it's about time someone did it because it's a story worth writing about, and I can tell it to you right now if you'd like, whatever you want to know, it won't take long, so get out your pen and notebook and let's get started. I was born in . . ."

The other men had gathered in a circle around us as I stood there utterly dumbfounded. They were catcalling and laughing. I wasn't sure what to say, let alone how to say it in Spanish. And then I understood what I had best articulate:

"Escuchame, escribo una historia solamente sobre las nueces de los arboles." (Listen, I'm just writing a story about the walnuts from the trees.) And I pointed at the trees around us.

"Ohhhh, las nueces de los arboles!" He smiled with delight, removing his hat and scratching his sweaty head.

"Ohhhh, las nueces de los arboles," the band sang and played in chorus.

Then he patted me on the back, stuck out his hand, and introduced himself as *el mejor dueño* (the big-shot owner). There was more jeering and commenting from the men about the kind of big-shot owner he was, and then they slowly scattered, leaving the two of us to get acquainted. He took me by the shoulder and led me over to the harvester and began to extol its virtues and to instruct me in how it worked. Finally, he insisted that I drive it. And after that he invited me to his home in Mexico to meet his

family, and basically to share with me everything in the world he possessed and loved.

Craig found Pedro and me driving on the harvester together way in the back of the grove. He and Pedro spoke for a few minutes. Like most farmers in California, Craig speaks Spanish, and when I joined him and Teddy on the ATV, he told me over his shoulder all about Pedro Garcia.

"He came to America in 1963 as one of the last Mexican workers in the Bracero Program, started by FDR during the Second World War.* He's an amazing man; he's like an uncle to our kids. He'll do anything for you. Great worker and role model. We feel very lucky to have him on staff. So you met the crew?"

"Yes I did."

"You know, I forgot to ask you if you can speak Spanish?"

"A little. Just enough to get in trouble."

"Well, I'm sure they appreciated whatever you had to say."

"Yes, I suppose they did."

Back at the house, Craig had lunch all made—homemade soup and bread, sliced turkey, cheese, tomatoes, lettuce, chips, and drinks. Again I felt totally spoiled, but tried hard to convince myself that I deserved it after such a trying morning, socially and linguistically speaking. He even served me. As we sat there spooning up the soup, he asked me if I would do him a favor. I

---

* According to Wikipedia, "The Bracero Program (named for the Spanish term *bracero*, 'strong-arm') was a series of laws and diplomatic agreements, initiated by an August 1942 exchange of diplomatic notes between the United States and Mexico, for the importation of temporary contract laborers from Mexico to the United States."

felt honored. He needed to drive Julie to the airport, but a good friend of his was due in from out of town at any minute, and he wanted me to wait for him and to feed him lunch when he got there, until Craig came back. He said it would take him less than an hour. He told me the man's name, which I didn't recognize, and then he told me the reason he was visiting: UC–Davis's College of Agriculture and Environmental Science (Craig's alma mater) was honoring the gentleman that evening with its "Friends of the College Award of Distinction." The man was A. G. Kawamura, former secretary of agriculture for the State of California, appointed by Arnold Schwarzenegger. Then Craig was up and on his way, helping Julie with her bags.

Sure enough, fifteen minutes later a car pulled slowly into the driveway. It was an odd-looking car, one I had never seen before, and I asked about it after A.G. and I and Teddy had introduced ourselves.

"It's a hydrogen fuel cell vehicle: the Honda FCX Clarity."

"I didn't realize Honda was making them?"

"Not only Honda but other automakers produce fuel cell cars."

"Really?" I didn't ask any more than that. *Only in California*, I thought to myself.

We sat on the porch, in the shade across from each other. I guessed A.G. to be in his late forties, dimples and big, fat, high cheekbones on a smooth round face, long black hair, thick and solid athletic build, about five foot eleven, but quite soft-spoken, and—*surprise, surprise*—self-effacing. I told him what my appointed task was, but he wouldn't take any lunch, just some juice, so I got him that and we sat there looking at each other, smiling.

"Craig tells me you're writing a book about agriculture."

"It's about harvest. I've gone around the country and participated in the harvest of about a dozen crops."

"Wonderful. Do you have any background in farming?"

"None. I garden. My grandfather was an immigrant who owned a fruits and vegetables distribution business back East. Growing up, we always had the freshest and the best fruits and vegetables to eat. So fruits and vegetables and nuts are kind of in my blood."

"My family of course emigrated from Japan to the United States. They were farmers . . ." He went on to tell me about his background in farming and about his family's thousand-acre farm. He said he had a very similar background to the Japanese-American writer David Mas Masumoto, who saved his family's heirloom peaches by writing a memoir, *Effigy for a Peach*. I wrote the author's name and the book down, as well as a couple of other interesting facts that A.G. shared. One was that California is the fourth-largest agricultural economy in the world. The other was that as a result of the previous winter California was sitting on a record snowpack up in the Sierra Nevada, promising ample supplies of water for the time being. At that point I asked him about the supply of water in California, because I had always heard that it was only a matter of time before the state suffered a major water crisis. But he assured me California had an aqueduct system second to none. The more we talked, the more apparent it became that A.G., like all the farmers I'd worked for, was a cockeyed optimist, and a Californian one at that, where the future was always quite literally very bright and very beautiful. But then we got into a bit of a discussion about GMOs. He started it!

I don't want to misrepresent myself here, or him for that matter. It wasn't a heated discussion, but it was a discussion. The upshot was this: he had no problem with GMOs per se. Then he told me a story about when he was the agriculture secretary and a delegation of Germans came to tour the state. They were very critical about Monsanto and its GMOs. But what he discovered was that it wasn't the GMOs they were opposed to as much as it was the fact that Monsanto held a monopoly on them. And that was his point, that GMOs per se are not the problem; the problem is the monopolistic intent of the corporation, specifically Monsanto. I didn't disagree with that argument, but I did say that I had a problem with GMOs per se because they weren't working. I told him about the lesson I had learned from Jim Gerritsen about the Bt-producing New Leaf potatoes and how it wasn't just a simple matter of the concentration of doses anymore. Everything was contaminated, right down to the DNA!

Fortunately for the both of us, Craig arrived back on the scene before we got too antagonistic. He bounded up the steps and onto the porch, his smile making all things copacetic. So with those two sitting and chatting on the porch, I excused myself and went off in search of my brother *trabajadores*, to rake and to harvest and to write about their nuts.

It took me a while to find them, way back in the endless grove. But this time, I found them all in the presence of another Jules Verne–like vehicle—the shaker. Low and compact, like the sweeper, it had a long, prehensile appendage in the front, about eight feet in length. The tight passenger compartment held just one man, behind a thin door. Other than the appendage, it looked a lot like a humvee, very heavy and low-riding, with thick steel all around, painted yellow. The driver, with a joystick, could grab the trunk of the tree with the viselike end of the appendage

and give it a shake. Whoever the driver was, he was an expert. He maneuvered the vehicle forward and backward, grabbed the trunk as if it were with his own hand, and gave it a vigorous shake (for about five seconds), before releasing his grip and backing away and finding the next tree. For five seconds the ground and the vehicle and the tree would vibrate, and walnuts would rain down every which way. It was an utterly efficient system, without a single nut left on a single branch. I was impressed by how much agitation those trees could handle, year after year. And to think that in the old days the *trabajadores* would climb up in the trees like monkeys and shake them with their legs, branch by branch.

Then Pedro insisted that I get into the vehicle and shake a tree. That was a near disaster. I tried to talk him out of it, but he insisted, so into the compartment I went. With the driver standing on the ledge, holding on to the roof, Pedro tried instructing me in Spanish, made nearly incomprehensible by the loud whine of the engine and drone of the hydraulic system, on the finer points of manipulating the joystick. It took a steady, delicate touch to move those snow-shovel-size tines into place and to grip the tree and shake it. Not being a video gamer, my less than fine motor skills just weren't equal to the task. That poor tree. I bumped it and bashed it and clobbered it and scraped it and shook it three times as long as it was accustomed. But I had done it, and I'm pretty sure Pedro felt a great deal more satisfaction than I did.

My daughter the comedian has a joke she uses whenever she does stand-up in her adopted city of Buffalo. It goes something like this: "You know, people always tell me I should move to San Francisco or Boston or Seattle, but I can't possibly live in a city that has better self-esteem than I have. That's why I love Buffalo." I was thinking of this joke as I found myself eating al fresco in the center of Winters, under a starry sky, with beautiful people all around me. After dinner, I strolled along the streets and looked into the shop windows. Again, it was like a movie set. There was a comestible store run and operated by Muslims all dressed in traditional white robes and kufis. A few doors down was a Mexican bodega with an elderly señora framed behind a colorful counter. Still farther down the road was a winery serving expensive bottles of local wines to fancily dressed young couples. Winters was perfectly polycultural; no one was excluded. Heck, it even had an opera house.

The opera house, aka the music hall, was on the second floor of an old building. At eye level along the walls in the first-floor hallway was a warmhearted display of photographs, all head shots, of local residents. Under the one-by-two-foot black-and-whites were neatly typed transcriptions of their personal histories—of how they came to Winters and what their lives there were like. Most were immigrants. All spoke glowingly of the beauty of the place and about the quality of their existence. I read them all. It was a terrific idea, whoever's it was, to display this sort of personal history as validation of your town.

Up in the old opera house was a guitar player from Texas, famous not so much for his highly stylized blues guitar playing as for his style of guitars—they were all made out of things he found, like washboards, cigar boxes, ax handles. Then I noticed Sean. Again he'd appeared out of nowhere. We sat and drank a few bottles of beer and listened to the music and talked about

Pedro Garcia. Sean said something that really took my breath away. He said that even before he spoke any Spanish at all, he understood everything that the man said to him. *There was something truly magical about Pedro.* Later we walked around Winters. At the end of the tour, I insisted on showing him the portraits and testimonials downstairs in the opera house. He knew all about them, but much to his credit he stood there and read them again in silence.

When he was done, I said: "Man, it's really touching to read these stories. First-generation Americans are the greatest Americans; they love, honor, and cherish it more than all the rest of us put together, and yet it blows my mind just how many chuckleheads out there want to deport them."

He nodded in agreement and said, "Yeah, but at least those kinds of people don't live in Winters."

The next morning I was put to work on the sorting line. I was given earplugs and thick rubber gloves, and the option to wear the white jumpsuit, but I declined. There were five of us on the line, pulling out walnuts that had any part of the green husk still intact after having gone through the steel brushes back inside the guts of the roaring machine. The conveyor took the freshly brushed walnuts from the sorter, up into the storage bins under the rafters above. It was contemplative work, as well as physical and rhythmic work. There was nothing asymmetrical about it, and I had no aches or pains or discomfort of any kind for as many hours as I stood there pulling out walnuts.

All morning I scanned the conveyor for signs of green, and even with four fast hands ahead of me, I had plenty of work at my end of the line. We took the unfinished walnuts and threw them into a big box behind us, where they would eventually be sent through the scrubber again to be dehusked. It was nearly impossible to talk over the drone of the machine, so any real communication with *los hombres* was impossible. It didn't matter, though, because on the line we were all one. I wasn't suspect, either. No, I was pretty sure Pedro had let them all know who and what I was all about.

And as I stood there on the assembly line, the lone gringo, my mind kept coming back to the idea of happiness. How contented they all seemed, and yet how proud at the same time. I had asked Craig about the statistics regarding Mexicans moving up the ladder, from farmworkers to farm owners. He sort of grimaced and said that this was one of the objectives of the Center for Land-Based Learning, to try to get that fire started, to get more workers to become owners. These men were paid well. Craig's foreman made $60,000 a year. So, perhaps, the incentive to go from worker to owner was not yet apparent.

But then I recalled the parable I'd read about the American investment banker and the Mexican fisherman:

> An American investment banker was at the pier of a small coastal Mexican village when a small boat with just one fisherman docked. Inside the small boat were several large yellowfin tuna. The American complimented the Mexican on the quality of his fish and asked how long it took to catch them.
>
> The fisherman replied, "Only a little while."
>
> The American then asked him, "Why don't you stay out longer and catch more fish?"

To which the Mexican replied, "I have caught enough to support my family's immediate needs."

The American then asked, "But what do you do with the rest of your time?"

The Mexican fisherman said, "I sleep late, fish a little, play with my children, take siesta with my wife, Maria, and stroll into the village each evening, where I sip wine and play guitar with my amigos. I have a full and busy life."

The American scoffed, "I am a Harvard MBA and could help you. You should spend more time fishing, and with the proceeds buy a bigger boat, and with the proceeds from the bigger boat you could buy several boats. Eventually you would have a fleet of fishing boats, and instead of selling your catch to a middleman, you would sell directly to the processor, eventually opening your own cannery. You would control the product, processing, and distribution. You would need to leave this small coastal fishing village and move to Mexico City, then LA and eventually New York City, where you will run your expanding enterprise."

The Mexican fisherman asked, "But how long will this all take?"

To which the American replied, "Fifteen to twenty years."

"But what then?"

The American laughed and said that's the best part. "When the time is right you would announce an IPO and sell your company stock to the public and become very rich; you would make millions."

"Millions . . . Then what?"

The American said, "Well, then you would retire and move to a small coastal fishing village, where you would sleep late, fish a little, play with your kids, take siesta with your wife, and stroll to the village in the evenings, where you could sip wine and play your guitar with your amigos."

After lunch, I went and worked with Pedro Garcia and the gang on the other side of the road, raking walnuts. It was very hot, and the trees were smaller, with less of a canopy, because it was a newer grove. They were really earning their dollars that day. We took more breaks than usual, and it was the first time I saw the men eating the walnuts and drinking lots of water. I did the same. Sitting under a tree with the older man next to me, I asked him out of the blue: "Cómo se dice en español 'las personas que vienen y recogen las nueces no recogido, después'?" (How do you say in Spanish "the people who come and collect the uncollected nuts, afterwards"?)

"Espigadoras! Sí vienen. Son todas mujeres. El dueño les paga bien." (The gleaners. Yes, they come. They're all women. The owner pays them well.)

"*Espigadoras?*" (Gleaners.) What a great word. Instantly I decided that the sequel of my adventures would be called *The Gleaners (Las Espigadoras)*, and I would employ myself exclusively in that capacity throughout the country with my wife . . .

My time at the Sierra Orchard was up. I had grapes to harvest and no time to waste. It was a very melodramatic goodbye under the trees, with Pedro Garcia holding my hand for the longest time and eventually relinquishing it in order to give me all his numbers and addresses in the United States and south of the border. I invited him to my home in Nueva York, and he was just

tickled pink to have a new friend from that mythical area of the world. Back at the hacienda, Craig gifted me a huge sack of walnuts, along with invitations to return with the entire family any time, and additional encouragement to send him my creative daughters to help out with the Center for Land-Based Learning. With the smiling sun above, and Teddy and Craig there in the driveway smiling and waving at me, I exited, stage right, once again, toward one last harvest.

Grapes

Of Rock Stars, Movie Stars, and Petite Sirah

ENVY WINES ESTATE WINERY

CALISTOGA, CALIFORNIA

DEVILS GULCH RANCH

NICASIO, CALIFORNIA

OCTOBER 2011

An ill wind was blowing. I felt it in my bones, but I couldn't just stop and turn around because, like Dorothy and the gang, I had come such a long way already.

From Winters I headed back west toward the town of Glen Ellen, in the northern Sonoma Valley. Once there, I was scheduled to join a team of grape pickers. But unbeknownst to me, a freak storm had blown in off the Pacific the night before and caused extensive damage to many vineyards in that region. I had pulled off the highway at a wayside above the city of Vallejo to call and check to make sure everything was a go, and found out that it wasn't. I learned that everyone in the valley was scrambling to save what was left of their crops. They were in no mood to host some East Coast scribbler who didn't know the difference between Shiraz and Syrah (is there one?). So I had to move to Plan B, even though I didn't have a Plan B.

I got in touch with a fellow writer who lived in Marin County and knew a lot of farmers in Northern California—Lisa Hamilton, author of *Deeply Rooted*.* I told her what the situation was. She had heard about the storm and recommended several names of people who hadn't been affected. Or so she thought. I wrote them down, thanked her, and just started cold-calling. It didn't

---

* Lisa M. Hamilton, *Deeply Rooted: Unconventional Farmers in the Age of Agribusiness* (Berkeley, CA: Counterpoint, 2009).

take me too long to get in touch with a man who owned a ranch and winery up in the hills above San Rafael. Mark Pasternak, owner of the Devils Gulch Ranch, invited me over to help pick grapes the next morning at either 7 or 10:30 a.m. I chose the latter. I asked him about accommodations and he suggested I go into San Rafael, which was fifteen miles due south of him off Route 101. He said I could find whatever I needed there. Boy, he wasn't kidding.

Two miles from the wayside, I turned onto Route 37, a raised causeway that arcs along the north shore of San Pablo Bay like an eyebrow from Vallejo all the way to Route 101 just north of San Rafael. There is a Far Eastern aspect to the northern shore of San Pablo Bay, with the treeless, sea-level bottomlands running in all directions for endless miles, and steep mountains off in the distance, dark and final looking, marking the terminus of the North American continent. On the land between the road and the bay, for almost the entire drive, I was aware of heavy farm machinery in motion under billowing clouds of dust, harvesting some product, but I didn't know what.

Then I was there! Whoa! San Rafael is a party town! And it was Friday night! The main strip was hoppin'. People and cars were bumper to bumper and shoulder to shoulder in all directions. I was lucky enough to find a parking spot right in the center of it all. I pushed a couple of quarters into the meter, held my breath, and dove right in. As I walked along the thoroughfare, I was pleasantly surprised to find a small city replete with every kind of restaurant, fancy bar, sports bar, honky-tonk, art gallery, fitness center, café, gelateria, taqueria, etc. And I was determined to skylark through it all, which I did, finding myself a few hours later at an art opening in a fancy gallery clutching a

glass of Cabernet Sauvignon and listening to a superb jazz duo playing bass and guitar. (I didn't bother looking at the artwork. I certainly wasn't in any position to buy any of it.) I ended the night there in front of them, grooving all by myself. There was a little motel just up the road, and I was able to check in and hit the hay before midnight. It was a good thing I opted for the later harvest the next morning.

The Devils Gulch Ranch is fifteen miles from San Rafael, but those were like no fifteen miles I had ever experienced: up and down and over mountain passes and through redwood forests, along razor-edge curves, and through cattle ranches. I finally came to the idyllic little hamlet of Nicasio, which looked like something out of old Mexico with its whitewashed church and bell tower in the center of the town square. Very near the church, on the other side of the road, I had to punch in numbers at a private gate, and then cross a little wooden bridge. Beyond the bridge, I drove two more miles through ranchland, and up the dry mountain road until I came to another gate, where I punched in another set of numbers and climbed even higher. One more gate and I was in.

A very rugged dirt road led me past a vine-covered hillside with a windmill on top. Then down I dropped into a deep gulch, forest on one side and dry mountain chaparral on the other. I came upon a flock of ravens sitting on a wide, wooden fence next to the biggest, pinkest porker I'd ever seen in my life. It stood there against the brown hillside in obese bas-relief. I had to stop the car and get out to see if it wasn't some kind of piece of pork-roast modern artwork. As soon as my feet hit the road, the ravens took to the air, and the beast grunted and trundled a few feet down the hill. I got back in the car and continued down the ravine.

Finally I was there. The ranch was at the bottom of the gulch. It was a massive compound with three or four dark brown buildings, some running laterally and others vertically. There was a corral in the middle of it all. I could hear and smell horses, but not see any. I'll tell you another thing I couldn't see: a single entrance door. Chickens clucked. Dogs barked. Horses whinnied, but no one was home. I felt like I was roaming around in a dreamscape once again. I was going to try to call the guy, but my cell phone battery was dead. The forest ran down the hillsides and had the place surrounded. I half-expected to see a cougar slink out of the woods toward me. Up above, the sky was so blue it made my heart ache. I stood and watched the ravens soaring high overhead for a long time.

I don't know how many minutes passed. A couple of dogs came out to check on me, but I couldn't coax them over to lay a hand on them. Suddenly I was aware of a young woman, a Latina, perhaps twenty-five. She was very demure and soft-spoken and didn't seem concerned about a stranger standing out in the middle of the compound. She approached me without fear. I told her who I was and what I was there for. She made a phone call and then told me to go back up the hill to the vineyard, where the windmill was, and to wait there. He would come and find me.

So I did.

Ten minutes later a truck arrived. The driver got out, and we shook hands. Mark Pasternak wore a cowboy hat and jeans and a flannel shirt—the Marlboro Man with the classic rugged features and build. He was perhaps sixty, shorter than me. And like all farm folks I'd met, he wasted no time. He unlocked a gate in front of us that led up the hillside to the windmill, and in we went. About one hundred yards from the gate, we came upon a

work crew. They were loading big bins of grapes that looked shriveled and rancid and putrid. I went from one bin to the other, and they all looked the same. They, too, had had some of that bad weather, and it had caused a lot of damage. Mark was fairly preoccupied with the effects of the storm, and he bounded down the way to talk to one of his workers, leaving me there amid a half-dozen young Mexicans who were busy packing things up and gabbing among themselves. I overheard one of the young men near me ask another: "Es este hombre el nuevo dueño?" (Is this guy the new owner?)

I said, "No."

They both turned and shot me a look. It wasn't a friendly look. It was a look that I recognized: it was the same one I'd seen long ago in my days as a professional boxer when I'd opened the wrong door and walked into my opponent's dressing room. If I wanted to join them in the *vendange* (the harvest) then I had better be prepared to prove myself.

Mark talked for a long time as I stood trying to mind my own business. After about fifteen minutes he came back over to me and said, "They're done for the day. We really got hammered by that squall. We'll let everything dry out. Why don't you come back on Monday morning, seven o'clock. We'll start picking again then."

I was back in my car and driving up the dirt road, and beginning to think that maybe I wasn't meant to pick grapes. I certainly couldn't stick around until Monday because my flight back east was scheduled for Tuesday morning. If I was going to pick grapes, it was now or never.

In five minutes' time I was driving back through the first gate, right into little Nicasio, which seemed to draw me into its center like a vortex. And like all things in California, it was straight out of a movie set (because it probably was one). Again, it presented a little like a piece of old Mexico as I drove to the town center, which featured a long ranch-style building with a porch. There was a post office on one end and a restaurant on the other, with a bodega in the middle. I parked in front of the post office and hopped out of the car. There was a group of about a dozen cyclists standing there, taking a break from their ride, all brightly clad in spandex and helmetry, straddling their ten-speeds and sucking on water bottles. I needed to make more phone calls, but my cell phone was still out of juice. I thought I would see if the post office had any outlets I could plug the charger into, and sure enough, it did. While it charged, I went back onto the porch to read for a while, with the warm sun on my face.

When the phone was charged, I called Lisa Hamilton again and told her my predicament. She suggested I try calling people back up in the Napa Valley, farther east, and she gave me a few names. I tried two; the second was a go. The owner of the Envy Wines—another Mark, Mark Carter—in Calistoga, the last town in the Napa Valley, invited me up. He said they were doing some picking later that day and early the next, and everything was set to go. I told him where I was and that I had to get a little lunch but I'd probably be there in a couple of hours. Soon I was back on Route 37, driving east this time.

I knew things weren't right. I could feel it in the pit of my stomach. Sierra Orchards had been a dream come true. It would have been the perfect, high-note ending, but life doesn't always shake

out in niftiness, and I was just bullheaded enough to ride it to the finish.

And ride it I did. Route 29 is a two-lane highway that runs through the heart of the Napa Valley. That Saturday afternoon the road was a frickin' parking lot the entire way. The huge wineries I passed looked like theme parks. What am I saying—they are theme parks! At one point, somewhere north of the city of Napa, I happened to glance in my rearview mirror at the car behind me. It was a giant Mercedes; the peace sign hood ornament looked three times larger than normal in my mirror. There were two ladies seated inside, and the driver was wearing one of those fancy, wide-brimmed hats like at the Kentucky Derby. She had on big Hollywood sunglasses, too, and was talking a blue streak to the poor woman in the passenger seat. Every time I looked she was flapping that mouth. But then I happened to look at the woman she was talking to. It was Teri Hatcher or Teri Hatcher's twin sister. And that little brush with fame kept me behind the wheel and somewhat focused for a good hour or more.

Who knows what time it was when I arrived. All I know is that I was completely frazzled and disgusted and fed up with the traffic, with Teri Hatcher and her wide-mouthed friend, and the entire vineyard vibe by the time I staggered into the great faux Italian villa. Now I'm sure there are people who will dispute this, but it seemed to me, judging by the hills closing in on either side of Route 29, and how the highway divides right there, that Envy Wines of Calistoga is the very last winery in the entire Napa Valley. Not that it means anything, but it felt symbolic at the time.

Inside I asked to speak to the owner, Mark, but he had gone somewhere. So I waited and waited and had to pass the time by choking down a few glasses of Cabernet Sauvignon, Petite Sirah, and Merlot. I stood there at the counter observing the well-dressed and seemingly well-heeled folks who, every time I looked, stood there in regal poses as if they were on a photo shoot for GQ or *Vanity Fair.* On my side of the bar, I was doing a pretty fair imitation of one of the Stooges at a wine tasting—"Hey Mo, pass the Petite Sarah . . ."

It's all a little muddled in my mind what happened next, but Mark finally arrived and I was kind of pushy or insistent, I should say, about getting out there in the vineyard and picking grapes. There were customers or clients he was busy with, but he was trying to be accommodating to all of us, even though he wasn't taking me out to the grapevines. Finally, when his clients or customers resumed their poses, Mark, full of effusive energy and a galloping stride, grabbed me by the arm and led me out to his vineyard to let me join the grape pickers. But we couldn't find them anywhere. They had finished for the day.

So he said the heck with it and showed me himself. He had these big muscular hands like a butcher's, and he started to rip the grape leaves off the vine, "leafing," he called it, not touching the grapes themselves, but just the leaves. He explained that he was uncovering the bunches to expose them to the sun for one last sun-baking that would pull out the sugars before harvest. This was done the day of or the night before harvest. And he let me try it. I felt like I was destroying the grapevines as I ripped into the leaves with my bare hands, breaking stems as well. But this was how it was done. He assured me it didn't hurt the plant. And it was a special sight, those exposed grapes hanging down

like bulls' balls in the shimmering sunlight. And all around, the most perfect pastoral setting pulsated. Green mountains, tinkling rivulets, soaring birds of prey, tall eucalyptus and cypress trees abounded. And the tilth! He had me take a handful of it and put my nose in it. Like espresso grounds combined with peat moss and potting soil. It was the richest, lushest humus I'd ever felt. I was tempted to eat it.

Mark told me he had eleven acres and made about fifteen hundred cases of wine each year. He hasn't grown his business, either. "By staying the same size, it's steady as she goes, you know. But I prefer to make a great handmade wine rather than a ton of okay wine. You tried my wines. What did you think?"

"Fabulous." (Not that I would know.)

Back inside, he introduced me to Mike Smith, his winemaker. I didn't understand. I thought Mark was the winemaker. No, he was the farmer and the owner; Mike was the viticulturist, the oenologist, the vintner. Hmm. Things were starting to zoom way over my head. But we all agreed that I should meet Mike the next morning around 5 a.m. at a famous vineyard south of there to participate in the harvesting of the grapes (*hiccup*!).

I found Calistoga to be a sweet and delectable little town with all the fixings, but far too precious for my tastes. Besides, they wanted $250 for a motel room. I laughed in the poor front desk clerk's face. After a great Mexican meal, I found a nice bar stool at a low-key place, watched a late-night football game, returned to the rented car parked on a side street away from the streetlights, tipped the seat back, and didn't wake up till the cell phone rang at 5 a.m.

Dr. Crane Vineyard is famous in the winemaking world. It is the most coveted of all Napa Valley vineyards, and according to my host Mike Smith, the best wine produced in the Napa Valley comes out of that little vineyard at the foot of a hill. Of course I couldn't see anything in the dark as forklifts whizzed back and forth in front of my curled-up toes. All I could see were the lights of those forklifts and two powerful beacons beaming down on a small section of the vineyard from a cherry picker, as if a boxing match were getting ready to start. There was a crew out there picking grapes under the light as I stood next to Mike Smith, Envy's winemaker, vintner, oenologist, and viticulturist.

Mike had been gracious enough to meet me out on the road and lead me to the enviable vineyard. He was very pointed in warning me about the forklifts, cautioning me to keep back inside the rows of grapes where I was sure to be safe; he didn't want to be responsible for another victim hit by the roaring vehicles. From the moment I entered the vineyard, I could feel a frantic, off-the-charts energy, half-desperate and half-insane, like being on the floor of the New York Stock Exchange or the Chicago Mercantile or down inside a Bolivian open-pit gold mine. The smell of diesel was ubiquitous and the roaring engines and the bright lights overhead pushed my pulse ever higher.

He began by telling me that the grape pickers I was going to join were a well-known and well-respected team run by a middle-aged woman known throughout the area who'd been part of the scene for many years. After that the lesson turned toward all things winey. I learned, for instance, that Mike produced the highest-scoring Cabernet Sauvignon ever out of Dr. Crane. (He never referred to the place as a vineyard, only to the man himself, the original owner of the land.) It didn't take me long to

intuit that Mike was a rock star winemaker. He even referred to winemakers as rock stars but tried to convince me he wasn't one of them. (In the Napa Valley, only rock star winemakers need apply.) Mike reminded me of a lot of people I know whose primary motivation in life is adrenaline rush. They climb mountains, scale rock faces, surf big waves, paddle over waterfalls, sail across oceans, swim with the sharks.

Mike's five-minute lesson grew into an hour-long treatise, not only about winemaking, but about wine-being, wine-seeing, and wine-doing. The problem was, I just wanted to pick grapes.

Nonetheless, I learned about Cab (Cabernet Sauvignon) and about the Clone 337, the original grape responsible for this venerable brew. I learned about the thirty-three Grands Crus of Burgundy, those super-special vineyards in central France that grow the world's greatest wines, and against whom all other vineyards and wines are compared. (*Yawwwwn.* I was pretty tired from the night before. Sleeping in a car isn't like normal sleeping.) But they're all in Burgundy, these Grands Crus, and America doesn't have a paradigm against which to compare, only Dr. Crane, which is why this place was hallowed ground. I learned about the scoring process of wine evaluators. A 100 on any variety would make a winemaker an instant superstar, with an income potential of a top-recruited NFL quarterback, Major League Baseball pitcher, or NBA forward. As far as Napa versus Sonoma . . . (*YAAAAWWWWWN.* He wasn't picking up on it.) Sonoma makes the best Pinot Noir and Chardonnay, while Napa makes the best Cabernet Sauvignon and Petite Sirah (wonder why *petite?*). Then he was talking about some hundred-year-old, thousand-dollar bottle of Cab that a rich friend of his had insisted he taste, and he was extolling its extraordinary virtues despite its age when I finally had to blurt out, "Okay, I've just got two things to say: I used to live in Belgium, where they make the world's best beer. And I tasted their very best beer, which is produced by monks up in the hills and bottled in traditional corked bottles, just like wine, and I think I paid five bucks for a .750 liter. Why does wine cost that much more?"

A great laugh exploded out of his mouth, similar to the one that erupted from my mouth when the hotel clerk quoted me a price of $250 for a room.

"Are you kidding me? Beer? Ha ha ah ha ha. I can reproduce that. That's like making a loaf of bread. You mess it up; you just do it over again. Winemaking has so many variables, so many nuances and subtleties and intangibles. The weather, the vineyard, the grapes, the sugar content, even the grape pickers . . . the . . . the . . ."

"Diesel fuel?" I asked.

"Yeah, that too."

"Which brings me to my next question: is there any chance I'm going to get to pick some grapes?"

"Oh yeah, sure, sure."

A moment later he was leading me toward the light. And the closer we got, the more informed I became. And when we were finally standing within sight of the crew, I froze in my tracks. There were six of them, three to a side. They had knives that flashed and sparkled. We stood in silence watching them go by. They came toward us, all lit up and brightly colored and dark-skinned, like a reenactment of a Diego Rivera mural. Their hands moved in blurs, like hummingbird wings, slicing and pulling, pulling and slicing, the ripe grape clusters falling into boxes below their feet with methodical thuds. They didn't move, really:

they flowed, as if they were standing on an escalator moving laterally along the row. One moment they were there to our left, and the next moment they were there to our right. And as they passed, a voice inside my head whispered sarcastically: *Silly rabbit! Grapes are for wine. They're not a crop: they're a precious metal, a rare gemstone, a fabulous weekend for Teri Hatcher and friends . . .*

I didn't pick a single grape.

AFTERWORD

Back in 1987, Jared Diamond wrote an essay in which he claimed that the worst mistake in the history of the human race was agriculture.* There is irrefutable truth to that statement. Agriculture consumes more fossil fuels than any other industry; likewise it contributes more greenhouse gases than any other industry, which in turn exacerbates global warming. It's the number one cause of water, soil, and air pollution. It is directly responsible for the obesity epidemic in our society. It is also the reason we now continue to double our irrepressible species every forty years. Yet if we look more closely and analyze the situation, crop by crop, we could imagine agriculture as the greatest boon in the history of the human race.

The poet laureate of labor, Studs Terkel, described the nature of work in America in his masterpiece *Working* as "violence—to the spirit as well as to the body. . . . It is, above all (or beneath all), about daily humiliation." After harvesting crops down on the family farms, I felt none of those violations of the body or spirit (well, perhaps a little bit to my body). To the contrary, I felt renewed and inspired, and so, it seemed to me, did all the ex-

---

* Jared Diamond, "The Worst Mistake in the History of the Human Race," *Discover Magazine*, May 1987.

traordinary farmworkers alongside whom I picked and plucked and pulled. Likewise, I compare the bosses I had down on the farm to many of the obtuse superiors I have endured in my professional life: the farm bosses made me feel vital, productive, and necessary rather than diminished, denigrated, and replaceable. These farmers, virtually all of them, possess rare and unusual talents and qualities, the very same that we seek in all of our leaders. People like Julie Rawson, Jim Gerritsen, Roger LaBine, Craig McNamara, Lin Davis Stephens, and Bryce Stephens and all the rest would duly fill the roles of principal, reverend, chief, elder, justice, general, senator, president, father or mother, and friend. I trust these people utterly. They are all truly great American treasures.

After 9/11 and the subsequent Cheney/Bush years, I thought America had lost its meaning. Any remnants of Jeffersonian democracy, Whitmanesque optimism, King civil rights commitment, or even pilgrim pride had disappeared. In its place was an America where money and government, Congress and corporations, had joined together to flout and in some cases rewrite the rules, to chastise and ridicule justice, and to obviate and eliminate democracy altogether. Where schools and universities once taught character, quality, and inspiration, they now only taught math, science, and humanities. Where the only message coming from the idiot box was "Look at me, I'm f**kin' famous!" Where the words *jobs* and *terrorism* were code for the word *profits*. Where environmentalism and conservation and planetary stewardship were considered partisan politics. Where freedom of speech and freedom of religion and freedom of assembly were suspicious behaviors.

I started this journey in search of America's lost meaning.

Like all of my endeavors, I set out without any expectations. Perhaps it is more accurate to say that I assumed the worst. That's my nature. In my search for meaning, I expected to be horrified, humiliated, and disappointed at every turn. And, in fact, if that were the case, I planned to move out of the country. I hold two passports. I speak other languages. So do my children. I am not sentenced to life in America just because I was born here. I am a son of the earth; *sono uno figlio della terra, je suis un fils de la terre, soy un hijo de la tierra.* I can live in more than two dozen nations and enjoy the same rights and privileges and immunities as the citizens there do. For me, America is not about freedom; it is about choice. But I can tell you from my heart, worldly reader, that after traveling around this land and working on the farms and living with and getting to know the farmers, I have discovered new roots and new meaning to this land I call home. I have learned from these farmers and farmhands what it means to be truly wealthy—wealth of purpose, wealth of fecundity, wealth of sustainability, wealth of autonomy—and what human life is like beyond the intervention and mediation of systems and institutions. There are pockets of purity and self-sustaining sovereignties all over this sprawling nation of ours. And there is new optimism, new pride, and yes, even a new kind of democracy beginning to grow again. I have seen it with my own two eyes, squeezed it in my own two hands, and drunk it straight from the source. It may be *slow* and it may be a little dirty and a little impoverished, but it is steady and it is real. If, like me, you are looking to rediscover optimism and pride and democracy and meaning, then go home to the farms, or just go out into your backyards and dig into the soil, and you will find it all right there.

Thank you.

ACKNOWLEDGMENTS

I would like to thank the following people, without whose selfless generosity, professionalism, and élan *Harvest* never would have seen the light of day: Carl Lennertz, my friend and erstwhile editor, who planted the original seed that I eventually harvested. Tom Powers, my mentor and friend, whose sage advice continues to light the road ahead. My editor and creative twin, Michael Signorelli, whose eye for detail was unimpeachable. My literary agent, Helen Zimmermann, who provided keen and timely advice from start to finish. My wife, Mary, and my daughters, Katherine and Evelyn, for their help, humor, and sacrifices. My oldest friend, beloved consigliere, and ideal reader, Matt James, aka Mothballhead, whose divine intervention provided the keystones for the making of this book. My sister, Debbie Horan Ferrer, and my nephew Matthew Craig Ferrer, whose wondrous talents continue to grace my works. Author Lisa M. Hamilton, who had my back all the way through my misadventures, and who was always just a phone call away with names, places, and numbers. Tom and Kathy Miano, who got me hooked on blueberries. Phil Coturri, the premier organic viticulturist in Sonoma Valley, who was ready, willing, and able to let me participate in the *vendange*; unfortunately Mother Nature had other ideas. Kiaran Locy, at the California Avocado Commission, who did everything in her power to accommodate me. Sara Jackson Miller, for her curiosity, generosity, and lickety-split assistance. Tom Fels, who provided tons of help early on in my adventure. Leigh Kuwanwisiwma, director of the Cultural Preservation Office of the Hopi Tribe, who invited me to harvest his family's sacred blue corn, but global warming and the crows had other plans. Wendy Fink-Weber, director of communications at the Western Growers Association, who made all the right choices. Emmalea Garver Ernest, of the University of Delaware, for her help and persistence. Rena Basch at Locavorious, for her suggestions. "Coach" Bruce Geffen, for getting me up to the Ann Arbor area. My in-laws, Ray Buckley and Barbara Buckley, who provided valuable contact information, transportation, and habitation. My sister-in-law Kathy McCann, who helped facilitate efforts in the Northwoods. Jeff "Dig It" Andersen and the Florence Griswold Museum for their creative and enthusiastic support. Ted Morgan for his joie de vivre and his treasure trove of stories. And finally, all of the farmers and farmhands mentioned in the preceding pages, from whose hearts, homes, and fields I have harvested such sublime inspiration.

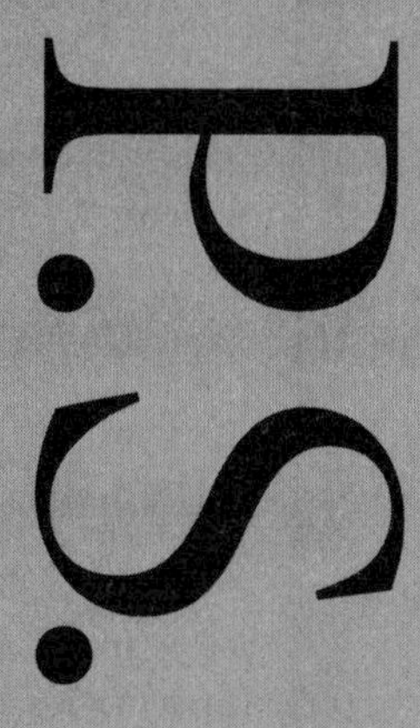

Insights,
Interviews
& More . . .

About the book

Read on

Photos from the Farm

On harvest with the header down at Stephens Farm, it's easy to understand what they meant by the phrase "seeing the elephant."

At Tantré Farm, "growing there in long, ornate lines, like a living text, the varied rows of herbs and vegetables, each plant a letter, each row a phrase, presented a transcendent form of communication."

Ready to sort blueberries with the Wiltse Farm team. From left to right, Dennis's mom Edna Wiltse (filling in for Flo, who exercised her right not to be photographed), Fred, Dennis, and Adriano.

In preparation for another CSA: pulling potatoes and loving every minute of it at the Many Hands Organic Farm.

Dan Trudel in his raspberry hoop house, literally dwarfed by the fruits of his own labor. "Isn't that amazing? *Huh? Isn't it!* Isn't it!?"

Photos from the Farm *(continued)*

The author dancing on manoomin, *with Roger LaBine looking on.*

The Great Gene Cranberry Sux Machine. That's Gene himself, shirtless, climbing up to the second story.

The crew takes a break from harvesting potatoes (and rocks) at Wood Prairie Farm, with Jim Gerritsen at the wheel.

"For five seconds the ground and the vehicle and the tree would vibrate, and walnuts would rain down every which way." The tree shaker of Sierra Orchards.

Winemaking is a serious business, as evidenced by the equipment at Envy Wines. Seriously.

Lamentations and Inspirations

I SENT A LIST OF QUESTIONS to each farmer with the request to answer all or none of them. I am delighted that many took the opportunity to respond. Here's what they had to say.

Julie Rawson, Many Hands Organic Farm, Barre, Massachusetts

What is your greatest frustration/ disappointment with farming or the farming life?

I wouldn't call it frustration or disappointment. More, it is anxiety about making budget. That has always been my major worry. That, coupled with a feeling of not being adequate to the task of running a farm. One must be good at so many varied things in order to be successful at farming—not only raising high-quality food, but managing machinery, staff, carpentry, marketing, budgeting, animals, plants, customers, etc. The other side of that is that farming is very challenging and there is never a dearth of knowledge, skill, and ingenuity needed to better practice this craft.

What is your greatest satisfaction/joy with farming or the farming life?

My greatest enjoyment is several-fold: Nothing beats being in the field with our staff and volunteers. We are a rather raucous crew and we have no holds barred. As Tyson says, "What is said in the field stays in the field." We dissect people's love lives, make fun of each other, talk politics and farm strategy, eat as we pick, sing songs, manage the small children and old people who are invariably on the crew, and have a great time as we speed through our work. I love a beautiful plant, and I love the laying hens that work all day long every day. And I love pigs that come woofing across the field, and I love walking barefoot in the fertile soil, and turning up earthworms when I am digging carrots. I love passing out nutritious food in people's CSA bags, and every new crop that comes in. I love the community that surrounds our farm. I loved raising our kids on the farm and that our grandkids are doing the same on their farm and our farm.

Lin Davis-Stephens, Stephens Farm, Jennings, Kansas

What is your greatest frustration/disappointment with farming or the farming life?

I, as an anthropologist among Native Americans, have observed suffering and loss and what is called "the Vanishing"; it has been expected among the farming communities as well. The family farm as a remnant of "American culture" is not only vanishing, but because of American politics and policy, it, too, is being extinguished.

What is your greatest satisfaction/joy with farming or the farming life?

Freedom, the sense of open country, the place to be close to God.

Do you make your living solely as a farmer or do you or your spouse have jobs outside of farming that help supplement your income?

My off-farm jobs and inheritance helped my children go to college and helped pay for some basics.

Do you have health insurance? Do you provide health insurance to your employees?

The farm used to help pay Grandma's insurance but not Bryce's, so my off-farm job pays for his.

Do you have a relevant story to share about the history of your farm or a significant event that is representative of your story/life on your farm?

I had four children, born at home. Crystal was born on the farm in the farmhouse; Demetria was born on the farm in the trailer house/mobile home. They both bonded with the place and the people, likely in daily dimensions and in dreams in ways I may not know. I remember my ▶

grandmother's house and the moods of the day when I spent my summers with her in eastern Kansas. I deeply sense those days of my childhood there; although I was born in a hospital, I did bond with her place. My son Bryan and daughter Roselin were born in the city but born in the home with midwife assistance and doctor backup. They have the common sense needed to live and thrive on the farm because their parents gave them self-respect and responsibilities for life and death carried from generations of farm/ranch customs.

My children were born at home and schooled at home. The farm life is an authentic educator. They participate fully in American society and are good team members with strong work ethic. The legacy of agricultural communities may still remain viable.

Deb Lentz and Richard Andres, Tantré Farm, Chelsea, Michigan

What is your greatest satisfaction/joy with farming or the farming life?

Deb's greatest joy is in the diversity and richness of this life of farming. This diversity naturally has to do with food variety, but it also has to do with the diversity of people in our community and what they bring into our lives. There is great satisfaction in knowing that we are providing food to our community that is sustainably rich in nutrients, flavor, and quality from the land that we nurture, but mostly in sharing this food, land, and sense of community with so many others.

Richard says that his greatest satisfaction is sitting down to lunch or supper with a group of people who have shared in the work and the harvest of the food that we grow. Also, there is satisfaction in watching plants and trees grow, mature, and bear fruit over the course of seasons and decades. He also likes the shared experience that we have with so many people and the joy of the harvest. It is important to be a member of a greater community where we are valued and depended upon for our contributions to healthy food, healthy ecosystems, and a healthy economy.

Do you have a relevant story to share about the history of your farm or a significant event that is representative of your story/life on your farm?

When Richard moved here, he had a job off farm to pay for the farm. He would come home at the end of a workday and continue working until dark. He worked seven days a week. That first spring Richard planted three small persimmon trees in the barren, hot, empty backyard. There were no trees and no outbuildings back there. Now, after we have lived here for eighteen years, there are two chicken coops, a CSA distribution center, a children's play area with a stick hut for imaginative play, an outside earth oven, a big red barn and packing shed, several other smaller outbuildings, and of course those thriving persimmon trees, which now provide shade, coolness, climbing opportunities, and tasty fruit in the fall. They are a place for families, farmworkers, and the CSA community to gather for CSA distribution and networking, picnics, weekly garden pizzas in the earth oven, flower arranging, preserving, many meals, and the sharing of lots of stories. Those persimmon trees are representative of our story at Tantré Farm . . . one of growth, connection, and community.

Can you give details about the development of your farm over the years and what you plan for the future?

Richard started this farm step by step, gradually building relationships with the neighborhood, the surrounding community, and the land. It started in 1993 with a few crops, such as potatoes, squash, and garlic, being distributed to a few stores in town. Over the years Richard continued to work full-time as a timber-frame carpenter and Deb taught elementary school during the school year with summers off. Then in 2001 Richard and Deb decided to start a CSA vegetable farm with thirty members. That summer they began working full-time on the farm, and their daughter, Ariana, was born. Since then, and as of 2011, the CSA has expanded to 350 members. It is now a thriving CSA farm with an emphasis on diversity of over eighty to one hundred different types of fruits, vegetables, and herbs, and distribution of produce has expanded to three farmers' markets, along with many wholesale accounts to local stores, restaurants, buying clubs, and once in a while even to schools and hospitals.

We started this farm by sharing the work with a few people who needed a place to stay and wanted to work outside. The details are really a story of community and interdependence. In other words, local roots and networking have been very important for the success of the farm, in order to develop a strong, supportive customer and CSA membership base. The importance is to emphasize that our farm is successful because of our friends, family, and neighbors, which shows quite literally it takes a village to grow a farm. Now we have an internship program, too, which attracts people of all ▶

ages and from all over the United States to learn, live, and work for a summer season on our farm.

For the future we will continue to search for the golden balance of sustainability for our family, friends, and community, and our particular ecosystem.

Do you have health insurance? Do you provide health insurance to your employees?

We have catastrophic health insurance, but can't afford to have full health care. No health insurance for employees at this time, although we have helped those who have been hurt with hospital bills with our own money, if they don't have insurance. We consider the physical work outside and our farm-fresh food to be our best preventative health care at this point in our lives.

What does Tantré mean, and why did you choose it as the name of your farm?

Tantré is the name of Thich Nhat Hanh's four-year-old niece who stayed with him in the Buddhist meditation center of Plum Village in the south of France in the late 1980s. He referenced his niece during a lecture on Buddhism and psychology. Nhat Hanh told about a time when he was meditating, and Tantré came to him because she was thirsty. He poured her a glass of unfiltered apple juice, but since it was cloudy she refused to drink it, set the glass of juice on the table, and ran outside to play. When she came back inside Tantré pointed to the glass of juice still sitting on the table that now had no sign of cloudiness and exclaimed, "Look, my apple juice is so clear. It's been meditating!" Nhat Hanh explained that even this young girl understood that meditation is about clarity and stillness, despite the fact that she had no formal training or education about it. Richard chose the name Tantré when he moved to this farm because after many years of searching for direction in his life, he finally settled on his childhood aspiration of becoming a farmer. This farm seemed to embody that very essence of meditation, one of clarity and stillness of mind and body, so it became "Tantré Farm."

Dennis Wiltse, Wiltse Farms, Constantia, New York

What is your greatest frustration/disappointment with farming or the farming life?

The weather. I live and die by the weather.

What is your greatest satisfaction/joy with farming or the farming life?

Watching the little kids come out here and frolic and play around and eat the berries. I also like to watch the way they try and rip me off, you know, putting rocks and water in the buckets.

Can you give details about what you plan for the future of your farm?

Cattle, going down the road. And I'd like to turn the extra land here into a small fruit farm with strawberries, blackberries, raspberries, some sweet corn, squash, and pumpkins. My pumpkins and squash did really well this past fall.

Do you make your living solely as a farmer or do you or your wife have jobs outside of farming that help supplement your income?

No, I don't. I do all kinds of work in the winter. I work as an auto mechanic for the road crews—the big orange trucks.

Do you have a philosophy, mentor, or successful system or practice that positively influences how and why you farm?

Just think that one day there won't be any more grocery stores, so I try to be self-sufficient. Last night I ate a great big steak, my own beef. I had some frozen squash, too. I eat like a king.

Are there any final thoughts or comments you'd like to make?

Just that people should support the local farmer. Drive around the countryside and buy fresh produce from the fruit and vegetable stands along the side of the road. Support the small farmers. They need the help. ▶

Dan Trudel, Ann's Raspberry Farm, Fredericktown, Ohio

What is your greatest frustration/disappointment with farming or the farming life?

I have been thinking about your first question for quite some time now and, quite frankly, I cannot come up with anything major that is of great frustration with farming, only minor things. Anybody involved in farming knows that a lot of "it" has to do with the weather and Mother Nature. Therefore, one should know that a great part of farming is out of our control. Just deal with it and adapt to it! If not, do something else!!!

Perhaps the greatest disappointment has been our son's refusal to do anything related to farming, and his reluctance to tell others that his mom and dad are involved with farming! So many people look upon farming as the bottom-ranked occupation, when in fact it is so humbling to work the land, not to mention the quality food we grow and produce. Still a lot of education to be done there.

What is your greatest satisfaction/joy with farming or the farming life?

Knowing that someone in his/her middle age can begin a farming career and be successful at it. But clearly, the greatest satisfaction has been the honor of bringing quality brussels sprouts and berries to the market and sharing fine products with people. Until three days ago, I would not have been able to fully bring this forth, but I just came back from a business trip doing market research in Wisconsin and Minnesota. I hadn't done any travel/work since last June and thus I have had the opportunity to spend the summer doing "farming" only. How great is it to work from home/field and have ample time to be at home with the wife and kids? Such a great blessing—not sure I am fully taking advantage of it, but! Beats the nine-to-five commute anytime!!

How long have you been a farmer and where were you born?

I would not consider myself a farmer yet, as I believe one should put at least ten years of farming under his/her belt. But yes, we do more than just backyard gardening. Just completed our seventh year, and I was proudly born in Montreal of French parents.

Ann Trudel, Ann's Raspberry Farm

I was born in Akron, Ohio. I graduated with a bachelor of arts in business administration/marketing in 1984 from the University of Akron. I met Daniel, from Montreal, Quebec, while on vacation in Daytona Beach, Florida, in 1983. We married in 1985 after a two-year long-distance love affair. Farming wasn't on our radar, but we longed to live in an Amish area after honeymooning near Berlin, Ohio. We fell in love with the rolling hills and charm of the country. I always kept a family garden and made raspberry jam with my sister and mother every year, picking the berries on a local farm. No one in my family was a professional farmer or in the food industry. We could never have dreamed of our growing success in farming. It has been very exciting. We never know who might call—even an insightful author who wants to write a book about harvesting!

It is important to us to educate people on buying local and teaching them about sustainable farming because we respect and admire our fellow farmers and the uniqueness of each farming venture. We have tours here and enjoy letting visitors taste our products made from the fruits and vegetables grown on our Certified Naturally Grown farm. It is important for everyone to know that farming is exciting and that many of the finest products available are being made in their own communities. We live in an Amish area of a small town, but several of our products have received the Good Food Awards in San Francisco, having been selected from almost one thousand products from around the country in a blind taste test. The Good Food Awards recognize the tastiest, most authentic and responsibly made products in the nation. One of our products, the Savory Brussels Sprout Gourmet Relish, has won two years in a row and is the one and only product in the nation to have been awarded a Good Food Award two years in a row! What fine, delicious products are passionately made in your own community? Get to know a farmer. You won't believe what you have been missing!

Roger LaBine, Lac Vieux Desert Band of Lake Superior Chippewa, Watersmeet, Michigan

What is your greatest frustration/disappointment with farming or the farming life?

Because ours is a restoration project and because we have only been open for harvesting for four years, there is a great lack of regulation and monitoring by state and local officials to help us protect the resource. ▶

Lamentations and Inspirations ***(continued)***

Anyone in Michigan can come and harvest, which obviously creates problems for us. Wisconsin, which shares Lac Vieux Desert, has strict regulations and licensing, so that only tribal members and affiliates have access to the wild rice beds. Michigan needs to enact the same regulations.

What is your greatest satisfaction/joy with farming or the farming life?

When my uncle Niigaanaash was diagnosed with cancer in 1994, I came back after being away for a long time. I feel the Creator called me back for a purpose. That purpose was to continue my uncle's work of restoring the *manoomin* harvest. The wild rice is our identity; it is who we are. Four years ago we opened the beds for harvesting to the people after a very long time. It is a wonderful feeling to be a part of that.

What do you envision for the future of* manoomin *on Lac Vieux Desert?

I sit on many committees. Presently, I am the chair of the Native Wild Rice Coalition, and we are partnering with the Great Lakes Lifeways Institute in order to bring back the native wild rice to the upper and lower peninsulas of Michigan. We hope to educate the public and the state agencies about how valuable this threatened resource is.

Brenda Cobb, Brenda Cobb Cranberries, Middleboro, Massachusetts

How long have you owned and operated your farm?

I've been operating this farm for three years. That is, I've raised three crops.

What is your greatest frustration/disappointment with farming or the farming life?

I think the only frustrating thing for me is that I'm not terribly mechanical, so I have to get help with equipment repairs. My farm is small enough that most things can be done by hand. I hand-crank the fertilizer, for instance.

Do you have a philosophy, mentor, or successful system or practice that positively influences how and why you farm?

This fits in with my lifestyle/philosophy of being in the moment and doing things in a thoughtful, connected way. Anytime I'm out there pulling weeds or digging ditches, I'm part of the big picture, and I can really see the plants. With the economy as it is, I can't make a living solely on the farm. I was thinking about this recently, and the fact that I have several part-time jobs, but that they are all things that I enjoy. Last year I got onto a local agritourism site and had many interested visitors. Sharing knowledge has always been a part of the cranberry industry. Local growers are always willing to come and check out any problem or answer questions with each other. And we are blessed with a fabulous cranberry experiment station with an excellent staff who are always available to us.

My dad still owns the property, and he has his ways and thoughts on how things should be done. I tend to lean more toward a softer, more environmentally sound and natural approach—as much as can be done without totally jeopardizing the crop. I was able to get a new irrigation system installed. Higher efficiency means more accurate use of any chemicals I need to apply.

Do you have health insurance? Do you provide health insurance to your employees?

For the past few years I've had insurance through the state. The only difficulty in being self-employed is affordable insurance. I have to say I am so grateful every day to be here. It's so exciting when the bog is in bloom and those first tiny berries appear. And then when they start to blush. Gorgeous.

Jim Gerritsen, Wood Prairie Farm, Bridgewater, Maine

What is your greatest frustration/disappointment with farming or the farming life?

We do not mind the long hours of farming, and we like the work. Without question the most difficult aspect of farming today is the financial challenges created by sixty years of the federal cheap-food policy along with the corporate concentration of agriculture, which has dramatically reduced markets and created a corporate stranglehold on agriculture. These factors deserve the greatest credit for preventing farm-gate prices from keeping pace with other sectors in the economy, liquidating farms, and bankrupting family farmers. For perspective, as a percent of income, the amount Americans now pay for their food is ▸

Lamentations and Inspirations ***(continued)***

less than one-half of what Europeans pay. The on-the-farm reality of this massive concocted economic inequity is an imbalance in the economy, a constant struggle for family farmers to pay the bills, and unrelenting difficulties in operating and capitalizing our farms on an ecologically and economically sustainable long-term basis. Family farmers want to grow good food and good seed, but we have been systematically constrained from doing so.

What is your greatest satisfaction/joy with farming or the farming life?

The reason we think farming is the best life is because our work is with nature and her incredible life forces. Every day our work involves care and we are able to observe and learn more about the web of life, which comprises the natural world. We are grateful that in our work we are immersed in and witness to new life: the sprouting of a seed, the birthing of a calf, the transformative creation of rich healthy soil which results from good husbandry of our land.

Can you give details about the development of your farm over the years and what you plan for the future?

We started out as a first-generation farm. We tried growing many different crops and it took us fifteen years to find our seed niche. We concluded that organic seed is a crop we liked to grow and had the aptitude and isolation for, and we were able to invent a direct-marketing mail-order business plan by which we could make our living. For the last twenty years we've been refining and improving our seed farming. The trends are good: over time our seed quality keeps increasing, our business is growing, and we are becoming more stable and secure. Now, as our children approach the age of making their major life decisions, we can provide for them and their families a good fertile organic farm and an established proven business which can provide their living farming, now and into the future.

Do you make your living solely as a farmer or do you or your wife have jobs outside of farming that help supplement your income?

All of our income comes from Wood Prairie Farm. We have no off-farm jobs.

Do you have health insurance? Do you provide health insurance to your employees?

For years we provided health insurance to our coworkers but eventually we became unable to keep up with the steady, significant yearly premium increases.

Can you explain the lawsuit against Monsanto in simple terms and what you hope will be the outcome?

OSGATA [Organic Seed Growers and Trade Association] v. Monsanto challenges the validity of Monsanto's transgenic (GMO) patents. Incredible as it may seem, Monsanto has standing, and should a family farmer become contaminated by Monsanto's transgenic seed, in addition to destroying the value of our organic crop through their invasion and trespass, we face the perverse prospect of a patent infringement lawsuit from Monsanto for "possessing" their transgenic technology and not having paid royalty on that "possession." Early on, our lawyer asked Monsanto for a legal covenant guaranteeing that they would not sue us in the case of a contamination incident, and Monsanto refused. Their refusal tells us they are intent on maintaining their option to pursue innocent farmers who become contaminated. Monsanto claims they have no plans to go after contaminated farmers, but why would we trust them? What if Monsanto changed its mind tomorrow? Since Monsanto won't provide the protection of a legal covenant we are forced to petition the court for justice and protection from their abuse of family farmers.

Additionally, we believe the U.S. Patent Office erred in granting to Monsanto their transgenic patents. We have prepared four distinct, independent, and self-standing legal arguments, which our lawyers will argue in court. Should we prevail with just one of our legal arguments, we win our case. Monsanto must defeat us on all four arguments for them to win.

We expect to win our case in federal district court. If Monsanto appeals, we believe the original ruling will be upheld in the Court of Appeals. After that, should Monsanto then appeal to the U.S. Supreme Court, we expect the Supreme Court will let stand the ruling of the Court of Appeals.

What would you change about agriculture if you were given the opportunity? ▸

Lamentations and Inspirations ***(continued)***

In their headlong rush toward greed and power, Big Ag has torn apart and discarded the "culture" from agriculture. Because family farmers have been forced from the land, the fabric of rural America has been devastated. To remedy this situation, I would rebuild the culture of agriculture and then reattach it to our American identity. I would reinvent a directness between the food a farmer grows and the food the country eats. I would refashion and rebuild the best aspects of our agrarian heritage: excellence based on hard work, sound husbandry, and decision-making biased toward long-term good over short-term profit; justice, fairness, and par exchange (parity) of farm goods; and shared values of security, self-reliance, and farming as an honored livelihood. Also included would be a universal recognition of our interconnectedness with one another and with the natural world. This ecological understanding would be manifested through a commonality of shared values: a rich and caring community based on respect between farmer and town, country and city, governed and those who govern.

Craig McNamara, Sierra Orchards, Winters, California

What is your greatest frustration/disappointment with farming or the farming life?

It is not really a frustration, but ensuring that California agriculture remains viable requires tremendous fortitude and vigilance on the part of those who believe in the culture of agriculture. This means being actively and proactively involved and engaged in creating a healthy, sustainable food supply for all Californians. Seven million Californians (almost 18 percent) are food insecure, which means that they do not know where their next meal is coming from. This is wrong and it is up to our generation of citizen farmers to correct this, and I know that we can.

What is your greatest satisfaction/joy with farming or the farming life?

I take great pleasure in knowing that I am producing a crop that is so healthy. Loaded with omega-3 fatty acids and chock-full of antioxidants, it is one of nature's greatest gifts! But my true joy is being a steward of the land. Knowing that the land we farm will remain long after I pass and realizing that our efforts to teach the next generation of decision makers about good land stewardship are successful is my greatest joy.

Can you give details about the development of your farm over the years and what you plan for the future?

In 1979 I began looking to buy a farm in our area. It was what is called a "seller's market," meaning that land prices were high and few farms were on the market. I spent a year going from one farm to the next, opening farm gates and knocking on farm doors, asking farmers if they were interested in selling their orchards. I finally found a farm that had been owned for several generations and sold the year before to a city slicker. After twelve months of grueling work, the city slicker had enough and wanted to sell. I bought the farm, never having been inside the farmhouse that we have happily occupied for the past thirty-two years. My wife was astounded that I would purchase the farm without going inside the house, but as I explained to her, I was buying our future, not our home!

Almost immediately after we made our first land purchase, an absentee landowner of a contiguous 650-acre farm called asking if I was interested in purchasing his farm. This was unheard-of at this time, a farm of this size coming onto the market. Knowing that I was financially strapped, I knew that I could not buy the entire ranch; however, I did purchase 120 acres of the best portion of it and alerted my neighbor about the balance of the ranch, which he immediately bought. I planted this acreage into walnuts the following year.

We struggled financially for the first ten years, barely being able to pay our loans. If it had not been for my father, who was my silent partner, we would not have made it financially during the time that we were improving the farming operation. We remained in debt for twenty-five years, until we sold a conservation easement in 2005 on the most beautiful piece of our property, forever ensuring that it would remain farmland and would never be developed into residential use. As a matter of fact, my father (long after he was my partner) questioned whether Julie and I were making the right decision in selling the easement, forever giving up our development rights, to which we said YES. We never wanted this farm to lose its agricultural base.

It was about ten years into farming the orchards that we made our first transition to organic production. Over the next five years we converted all of our acreage to organic. This decision was based on several factors, not the least of which was that we were living in ▶

the orchard and raising three children on the land. After every spray my wife would question what the hell we were doing spraying! And she was right.

There were also very significant financial reasons to convert to organic. The conventional market for walnuts was paying less and less, and payments were based on the size of the walnut. Because much of our production was from older trees, our nut size was declining and so was our income. We were able to correct this in the organic market and our profitability increased significantly.

We added the Center for Land-Based Learning to our operation in 1994. As a 501(c)(3) nonprofit, it has grown into a business incubator, employing twelve full-time people with an additional four to five significant partnerships with Audubon, Putah Creek Council, UC–Davis, and more. I do not at this point see our farming operation growing in size. However, as parts of the orchard age out (fifty to sixty years old), we will be replacing them.

Do you make your living solely as a farmer or do you or your wife have jobs outside of farming that help supplement your income?

My salary accounts for 85 percent of our overall income, with my wife's UC–Davis salary providing the balance. Her employer also provides us with health insurance, which is very important.

Do you have health insurance? Do you provide health insurance to your employees?

The University of California covers our family's health insurance. We have also invested in life insurance and retirement accounts. We provide and pay for comprehensive health insurance coverage for our employees. I have also established a 401(k) retirement account for our main employee and pay annual bonus packages of approximately 12 percent of their annual salary to all employees.

An Excerpt from *Seeds: One Man's Serendipitous Journey to Find the Trees That Inspired Famous American Writers*

Lincoln, Twain, Presley, and Faulkner

During the spring break of 2001, my wife, our two daughters, and I went on a vacation to the Gulf of Mexico. Destination: Dauphin Island, Alabama. We drove from our home in Wisconsin, covering more than a thousand miles of interstates and back roads. To break up the drive, we put together an itinerary of historical places to visit along the way.

First stop: Springfield, Illinois, and Abraham Lincoln's home. Originally just a cottage, the place was expanded by the Lincolns into a two-story, twelve-room house soon after they moved in. When we visited it, the saltbox colonial had an overabundance of creaking stairways, paisley wallpaper, crimson carpets, and primitive-looking furniture. And it smelled funny.

My youngest daughter, just seven at the time, was dazzled by it. I felt it lacked all "freedom of interior and exterior occupation," to borrow a phrase from Frank Lloyd Wright, but then again, the young Lincoln was not noted for his architectural contributions to the house, just his legendary prowess there with an adz.

In the living room, a photograph caught my eye: a picture taken in May of 1860 of Honest Abe standing out in front of the house next to a young basswood tree. ▸

An Excerpt from *Seeds: One Man's Serendipitous Journey to Find the Trees that Inspired Famous American Writers* *(continued)*

Coincidentally, there was a fully mature basswood in that same spot just outside the window.

"Say, is that the same tree as the one in the photograph?" I blurted out.

"I believe it is," the docent replied.

I felt a thrill run down my spine.

It was the perfect excuse to escape, so I left the family behind to continue the malodorous tour while I went outside to take a closer look at the ancient hardwood that had known Lincoln personally. There was nothing special about the tree—no patina-proud plaque pointing out its pedigree, no initials carved into the bark, no tattered rope swing. It looked like any other tree. On the ground and underfoot were scads of golden pea-size seeds. I don't know what possessed me, but I reached down and picked up a handful and jammed them into my pockets.

This tree had known one of the greatest and most complex figures in American history. Had Lincoln even leaned against it and pondered his future? Surely he must've dreamed under that tree, dreamed of a better life for his family, for his fellow citizens, black and white. Suddenly those seeds in my pocket from that touchstone felt like pennies from heaven.

Next stop: Hannibal, Missouri, on the western banks of the Mississippi River, just an hour and a half from Springfield. At Mark Twain's childhood home, we were disappointed to learn that we had missed the last guided tour of the day. I searched the yard for old trees and seeds. Nothing. But Hannibal itself was old and seedy, surprisingly untransformed by its onetime resident's fame. Later that evening, while my brood swam in the indoor pool at the hotel, I decided to have a look around town in the waning sunlight.

Because I have been a transient most of my life, I have a knack for bonding quickly with any given locale. I need only wander around a place for a little while to feel a keen sense of belonging. As a teacher, I've learned that someone's environment has as much to tell us about that person as does his or her friends and family. So, within the hour, *Ich bin ein Hannibaler.*

I came to the base of Cardiff Hill, that illustrious playground of Tom Sawyer and Huckleberry Finn. A rusty sign modestly boasted of the site's place in literary history. As I made my way up the steep incline along a narrow dirt path, I half expected two waifs to come bouncing out of the bushes in rolled-up dungarees, wooden swords in hand, battling make-believe pirates.

Standing there atop the hill, looking down at the broad, muscular river below, I suddenly realized I was breathing in Twain. In that sublime vista drenched in the heavy, ionized air of the river valley, his worldview revealed itself to me in one wet respiration. A crow called out behind me as if to clear its throat. I turned to follow it and found myself gazing upon, for as far as the eye could see, a proud stand of hardwoods—locusts, box elders, elms, maples, oaks—running north to south along the ridge behind me. These were the offspring (there were no ancients among them) of trees that had once watched little snot-nosed Sam Clemens at play. This time I had a pouch strapped around my waist; rummaging around the area, I gathered up what seeds I could find and deposited them in it.

Next stop: Memphis. The magic of Graceland is not found in the memorabilia sold at the gift shop, or in the heady opulence of His private plane, or in the less-than-grand entrance to the ersatz plantation, or in the cheesy sixties décor, or in the "chicken-fried" trivia, or in the Safari Room, or in The Hall of Fame, or in the jumpsuit shrine, or even in the divine bathroom where he expired. No! The magic of Graceland is found in the people's reaction to it.

So while my wife and kids listened to the guide, I people-watched behind dark sunglasses. In fact, I was so thoroughly entertained by the kaleidoscope of rapture that I'd almost forgotten about my new hobby: collecting seeds from the trees that once knew historically significant people. That is, until I found myself outside, between the Hall of Fame and the jumpsuit shrine. And there, on the lawn, scattered like tiny Elvis capes, was a sea of maple seeds. At first I was worried that the security folks might intervene, but no one paid me any mind as I knelt down and excitedly scooped up the little key-shaped pods and placed them in my pouch.

Sated, I wandered over to the line of people waiting for their turn to stand in front of the King's grave. It was while standing in that line, fingering my waist pouch as if it were filled with gold doubloons, that I had my epiphany: I would travel across America to gather the seeds from the trees of great Americans who had influenced my life or influenced the course of American history.

I would visit their hometowns in search of the trees that may have played a part in their early development and helped form their views. I'd look into their lives and works for references to trees. I would also seek out trees that had witnessed great historical events. ▸

An Excerpt from *Seeds: One Man's Serendipitous Journey to Find the Trees that Inspired Famous American Writers* (continued)

The names came flooding in. First, the champions of nature: Thoreau and Emerson, Carson and Muir. Then the novelists whose words were succor to me, as a student and then a teacher: Kerouac, Wharton, Shirley Jackson, Henry Miller, Vonnegut. The great poets, too, and American places: Gettysburg, Mount Vernon, Wounded Knee. The deluge of names and places cascaded through my brain for some time before it ebbed to a trickle.

Final stop: Oxford, Mississippi. I don't think there is a place on the surface of the planet that feels more uncomfortable to a native New Yorker than the Deep South. The air, the architecture, even the trees exude a sort of Yankee repellent. For some odd reason, it doesn't work in reverse; that is, Southern boys such as Truman Capote, Willie Morris, and even the late, great New York City chronicler Joseph Mitchell felt right at home in a Manhattan clam bar or on an East Hampton beach. I wonder why that is.

The city of Oxford is the quintessence of Southern gentility. At its center is a classic square around which sit antebellum structures made of fiery red brick trimmed out with white columns, iron-railed porches, and ornate roof moldings. We arrived, all of us jam-packed into a late-model Olds 88, during a torrential downpour. If the sun had been shining, I'm certain pedestrians would have halted in their tracks and kids at play in the square would have missed catching the ball as they all turned to watch the silver sedan with the Wisconsin license plates entering the scene.

Faulkner's home, which he named Rowan Oak, wasn't easy to find even with directions, but I spotted it at the end of a residential street: well hidden behind a thick grove of pines. There was a handwritten sign on the gate: "Closed for Repairs." Hell, we'd driven sixty miles out of our way to get here; no stupid sign was going to keep me out. It was still raining, so my family happily stayed behind in the car with the radio on while I hopped the fence and entered the property.

It was spooky in there. The majestic plantation-style mansion, with its giant white columns and large shuttered windows, eyed me suspiciously as I diffidently approached. There was no one around. Everything was still. The ample yard and the numerous living quarters of the once un-free help were well maintained, but there was a tragic, severe feeling to the place. The trees completely surrounding the property had an immuring edge to them. Illness lingered. I could imagine Faulkner's dark, rummy eyes watching my every move, his lips pursed around the mouthpiece of a bulbous-headed pipe, as I splashed around the outside of the house, peering in all the windows.

It had stopped raining and was misting; everything was steamy and gray and damp. My umbrella was of little use, so I pulled it closed and hung it from my belt. Completely drenched, my wet, tangled hair covering my face, I felt squalid. In a sudden act of exuberance, I sprinted across the lawn and did a feet-first slide up to the base of an old maple tree at the far end of the yard. And as I lay there, soaked head to toe and looking up into the matronly branches of the tree, my epiphany back at Graceland began to play through my head again: I would be sure to spend significant time on a Southern writers' tour—Eudora Welty, Carson McCullers, Richard Wright, Flannery O'Connor, Truman Capote, Harper Lee, and Tennessee Williams—and wouldn't miss a nonliterary hero of mine, Muhammad Ali.

I could feel the girls' impatience pulling me back through the dampness, so I gathered what maple seeds I could find and sprinted toward the car, vowing to revisit Faulkner.

I took my collection of famous tree seeds back to Wisconsin and planted them in our yard. A few sprouted, but most didn't make it, and the rabbits ate those that did. I managed to salvage a few saplings, and gave them out as presents to friends and family. But that's as far as it ever got.

A few more trips ensued: unplanned family events. A few years later, I visited Ellis Island for the first time. There stood ancient sycamores, still greeting all who stepped off the ferry. Imagine the millions who saw those trees at the start of their new lives? I collected pocketfuls of the seeds and stored them in my basement.

When my father passed away in the spring of 2005, I returned to my childhood neighborhood to drive by the house where I had grown up. I just wanted to make that connection before we put him into the ground. It was early spring, and the buds were beginning to bulge out on the trees. The blush of color to the scenery made everything look like a Seurat painting. Behind my old house, the tall, intertwining cherry trees from my youth were still there, but out in front, my great playmate, the red maple, was gone.

I drove down the block to the site of my elementary school. The building had been torn down decades before and turned into a small park, but behind it stood the same grove of hardwood trees I used to play among at recess. I parked the car and entered the four-acre woods. Nothing had changed but me. Standing there beholding the ▸

An Excerpt from *Seeds: One Man's Serendipitous Journey to Find the Trees that Inspired Famous American Writers* *(continued)*

same wonderful trees of my childhood, I felt a glow of belonging, of embrace. I remained in those woods for a long time. When I returned to the car, my heart was full to overflowing with the seeds idea once again.

That was four years ago. Since then, we'd moved from Wisconsin to upstate New York for my new teaching job. I compiled a list of dozens of great American writers whose homes I wanted to visit. Friends and colleagues, upon hearing of my idea, urged more names upon me, and I happily, if also anxiously, added them to my notebook. How would I possibly find the time to make these trips? No matter. I had to take action, to take the first step.

On an unseasonably warm day in March, I set out from my new home on the southern shores of Lake Ontario to collect seeds. Unlike John Chapman, a.k.a. Johnny Appleseed, I had no business plan and no gospel; and I would be taking seeds, not giving them. But like him, I was on a mission. I would start nearby and work my way out: short trips, then long trips out West, down South, and over to New England.

According to season and location, be it during summer vacations or on long weekends, I would go with family and friends or on my own. But bit by bit, I would gather the seeds, bring them home, and grow them, and then tell my family and friends the stories of the trees from which the seeds came and the lives and literature they touched.